Human Growth
after Birth

DAVID SINCLAIR

M.A., M.D., D.Sc., F.R.C.S.E

Director of Postgraduate Medical Education,
Queen Elizabeth II Medical Centre,
Western Australia
Formerly Regius Professor of Anatomy,
University of Aberdeen

THIRD EDITION

LONDON
OXFORD UNIVERSITY PRESS
NEW YORK   TORONTO

L ML K

Oxford University Press, Walton Street, Oxford OX2 6DP

OXFORD LONDON GLASGOW
NEW YORK TORONTO MELBOURNE WELLINGTON
IBADAN NAIROBI DAR ES SALAAM LUSAKA CAPE TOWN
KUALA LUMPUR SINGAPORE JAKARTA HONG KONG TOKYO
DELHI BOMBAY CALCUTTA MADRAS KARACHI

ISBN 0 19 263329 5

© *Oxford University Press 1969, 1973, 1978.*

First edition 1969
Second edition 1973
Third edition 1978

Typeset by the Universities Press, Belfast
Printed in Great Britain by
Richard Clay (The Chaucer Press) Ltd,
Bungay, Suffolk

Contents

CONTENTS

CONTENTS

Preface to the Third Edition

In preparing the third edition I have been very conscious of the exponentially increasing price of textbooks, and I have looked hard at any addition I have made. However, some additions have been inevitable, partly because of our increasing knowledge of growth processes and partly because of criticisms that certain parts of the second edition did not contain enough information. Several chapters are thus a little longer, and the book has swollen somewhat. Three new figures and two new plates have been introduced, and two of the old figures and two plates have been discarded.

The general character of the book is unchanged. As before, it is intended to give an elementary account of human growth for medical students, their colleagues in the paramedical disciplines, and students and teachers of human biology. I hope it will continue to be found useful as an introduction to an immensely complex and fascinating field of study.

Perth   D. S.
1978

1

Nature of Growth

*'Growth is a word, a term, a notion,
covering a variety of diverse and
complex phenomena'*

PAUL WEISS

DEFINITION

Everyone understands in a general way what is meant by growth.
We have all grown, and think we can speak from experience.
Great store is set by parents on seeing how their children are
growing, and most readers of this book will remember being
stood up against a convenient door to be measured by their
mothers. Nevertheless, very few parents have much idea of what
is going on in the child they are measuring, or how his progress in
this direction can be related to that of other children.

The fact that growth is such a commonplace experience has
perhaps hindered proper investigation of the factors involved,
and it is only in the last thirty years or so that we have begun to
realize just what a complicated process it is. The *British Medical
Dictionary* defines growth in the following terms: 'The progressive development of a living being or part of an organism from its
earliest stage to maturity, including the attendant increases in
size.' If we look in the same dictionary for the definition of
'development', we find 'The series of changes by which the
individual embryo becomes a mature organism'.

These definitions tell us a good deal. Notice first of all that
growth involves a series of *changes*, not just the addition of
material to achieve an increase in size. These changes include a
differentiation of various parts of the body to perform different
functions, and also alterations in the form of the body as a whole

as well as in the form of its individual organs and systems. Material such as bone and fat is added, but it is perhaps less obvious that other material is actually subtracted. For example, the thymus gland, which is large and prominent during childhood, gradually degenerates and is absorbed, until in the adult there is little trace of it left. In the same way vascular channels which were needed by the embryo but serve no useful purpose after birth become converted into fibrous functionless vestiges.

Growth thus involves the incidental destruction and death of cells and tissues. Such 'premature' death is sometimes said to be 'programmed', since it forms part of the programme of normal growth. It may occur as early in development as the blastocyst stage.

Growth may also involve substitution, for example, when cartilage is converted into bone, or when the permanent teeth replace the temporary ones. Less drastic are the processes of alteration and modification which are integral features of growth, such as the secondary sex changes in the shape of the bony skeleton. Similar modifications take place on the microscopic scale: for example, the cells through which filtration takes place in the kidney are cubical in shape at birth, but later become flattened to facilitate the passage of material through them.

Secondly, the definitions we are considering distinguish between the growth of the whole 'living being' and the growth of one or more of its parts, and this is important because not all parts of the human body grow at the same rate. Nor do they all stop growing simultaneously. Further, growth of one part may be controlled by the activity of another (such as the endocrine system), and the degree of control will depend on the stage of development reached by the controlling part. It is therefore not to be expected that the body will retain the same proportions throughout growth, or that the relative weights and sizes of given tissues and organs should remain constant. This differential growth of parts of the body also implies movement of one part relative to another. For example, because of the relatively very small pelvis of the baby [p. 72], the bladder is an abdominal organ in early childhood; as the pelvis grows, so the bladder sinks down into it.

The third thing to notice about our definitions is that they are incomplete, for they omit the fact that growth does not cease when maturity is attained. We are all familiar with the continued growth of our skin, nails, and hair, but there are also other less obvious forms of growth which continue throughout adult life. For example, the lining of the alimentary tract is constantly being renewed, and in nearly every tissue and organ there is a recurring cycle of growth, death, and replacement. Finally, there are the processes of disordered growth which result in various forms of tumours in the adult as well as in the child.

Two more points remain concerning definitions. The first is that growth is a continuous process extending from the time the ovum is fertilized until death. The most spectacular phase of growth occurs before birth, and with this we are not directly concerned in this book, though we shall find it necessary to refer to certain features of the embryonic period in order to understand the growth of the child. The second point relates to the use of the words 'growth' and 'development'. There is a tendency nowadays to use 'growth' with the restricted meaning of anatomical and physiological changes, whereas 'development' is used in relation to the emergence of psychological attributes, ideas, and understanding, as well as the acquisition of motor and sensory skills. In this sense we shall not discuss 'development', except in a very brief and superficial manner.

PROCESSES OF GROWTH

The human body, like those of other animals, consists of cells and intercellular matrix, and both these grow—the former in size and number, and the latter in amount. When the component cells of a tissue or organ increase in number by dividing, growth is said to be *multiplicative*: when they increase in size, growth is *auxetic*. When the non-living structural material between the cells is increased in amount, the resultant growth is called *accretionary*.

A distinction must also be made between *interstitial* and *appositional* growth. In the former the tissue grows uniformly throughout its mass, and in the latter the new material is added to the surface of the existing substance [FIG. 1].

Cell Division and Differentiation

It has been estimated that the number of cells in the human adult is of the order of 10^{14}, and all these are ultimately derived from

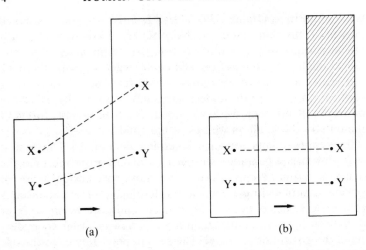

FIG. 1. Interstitial (a) and appositional (b) growth. In interstitial growth new material is laid down more or less evenly throughout the block of tissue, so that the distance between the cells X and Y increases as growth takes place. In appositional growth the new material is simply added on top of the old, so that the distance between X and Y remains constant.

the single fertilized ovum. This may seem a very large number, but in fact only some forty-five generations of cell division would be sufficient to attain it.

The ovum is a large cell [PLATE 1(a)] and its initial division into two as a result of penetration by the spermatozoon is not preceded by any increase in the amount of protoplasm it contains. The two daughter cells thus each contain about half the material of the parent cell. After a few generations of division, however, the cells do not divide until after a phase of enlargement during which they synthesize protoplasm and absorb water, and this sequence is followed by all subsequent generations. The manufacture of protoplasm is naturally dependent on a supply of suitable foodstuffs, in particular amino acids, which can be built up into protein and other substances. In the early stages of growth the embryo has a very high water content, but this gradually falls from about 98 per cent to approximately 65 per cent in the adult body (p. 38).

It is also worth noting that cell division in the first couple of generations occurs more or less simultaneously in every cell, but that after the cells begin growing before dividing the timing becomes staggered, so that in a given tissue only a small proportion of the cells present engage in division at the same time. By this means cellular multiplication is converted into a steady and gradual process instead of proceeding by fits and starts. How the staggered timing is controlled—that is, why certain groups of cells should be stimulated to divide while others of apparently equal maturity have to wait some time longer—is quite unknown, though it has been suggested that the peripheral nervous system could play a part [cf. p. 143].

There is a limit to the size a cell can attain, and this limit is set by purely physical and chemical factors, which in turn depend on mathematics. As a spherical cell grows, its volume increases with the cube of the radius, but its surface area only increases with the square of the radius. Now the volume of the cell determines its biochemical activity, since metabolism goes on all through the substance of the cell, but all the materials necessary for this activity must pass in and out through the surface membrane. As the cell grows, therefore, its biochemical function becomes more and more restricted through lack of raw materials and energy, and eventually some resulting chemical stimulus causes it to divide. What this stimulus might be is again unknown: it appears that cell size is not the only factor, since amoebae can be induced to divide when they are well below the weight at which division normally occurs. There is a limit to what can be done experimentally, however, and an amoeba less than one-third of the normal dividing weight will not divide whatever encouragement it is given. It appears that the stimulus to division originates in the cytoplasm rather than in the nucleus, and it is interesting that in multinucleated cells all the nuclei enter mitosis synchronously.

The restriction on size of individual cells can be to some extent overcome if the cell alters its spherical shape by elongating (nerve cells), flattening (epithelial cells), or folding up its surface membrane (intestinal cells), for in all these ways surface area can be increased without much change in volume. Cells which have to exist in closely packed tissues do not have much option regarding the shape they adopt, since again the laws of physics come into operation, and they tend to take up a shape which partitions the space available with a maximum economy of surface area. Lord

Kelvin showed that an assembly of regular 14-sided figures (tetrakaidekahedra) fits these conditions best, and this is the form which is seen in the interior of a mass of soap bubbles. A similar arrangement of rather more irregular 14-sided figures is found in animal tissues [Fig. 2].

Another factor which may play a part in determining the time at which a cell must divide is the ratio between the size of the cell as a whole and the size of its nucleus. The nucleus consists mostly of chromatin, and does not increase in size with growth at the same rate as the cytoplasm. The surface area of the nucleus, through which all interchange between the nucleus and the cytoplasm must take place, thus becomes inadequate to allow proper control of the cytoplasm by the nucleus, and it has been suggested that cells divide in an attempt to keep a fairly constant ratio between the amounts of the material in the nucleus and in the cytoplasm. Just as a cell can partly escape from the mathematical limitations to its growth by altering its shape, so the nucleus can allow some further increase in cytoplasm by becoming flattened or lobulated, though again there is a limit to the scope which such manoeuvres allow. A polyploid cell, in which the nucleus contains three or more times the haploid number of chromosomes [Fig. 3], can apparently accommodate an amount of cytoplasm which is more or less in proportion to the amount of nuclear material. Such cells are not common in the human body, but do occur, especially in the liver.

It will now be clear that the processes of growth and cell

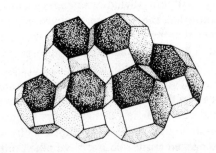

Fig. 2. Irregular tetrakaidekahedra. A regular tetrakaidekahedron has three pairs of equal and opposite quadrilateral faces and four pairs of equal and opposite hexagonal faces.

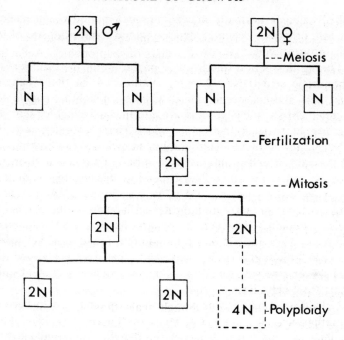

FIG. 3. Chromosome numbers (highly schematic). During the maturation of the male and female sex cells the diploid number of chromosomes in the nucleus (2N) is reduced to the haploid number (N) because the chromosomes segregate, in the process of *meiosis*, into two groups, one of which passes into each daughter cell (In man, N = 23.)

At fertilization, the diploid number is restored, and the offspring of the fertilized ovum maintain the diploid number because at the time of division (*mitosis*) each chromosome splits into two halves, one of which passes into each daughter cell. Occasionally, however, somatic cells with 3N, 4N, 5N . . . etc. chromosomes are found, and such cells are said to be polyploid.

Even-numbered polyploidy (e.g. 4N, 8N) can be produced by the chromosomes dividing preparatory to a mitosis which does not take place. Odd-numbered polyploidy, which is much less common, may be due to anomalies in the nuclear spindle at the time of mitosis.

division are complementary and interdependent. Growth of a cell beyond a certain size is impossible, and further growth of the tissue of which the cell is a part involves a temporary reduction of the size of the cell by the process of division. In the same way, cell division is normally dependent on a period of antecedent growth, otherwise repeated division would reduce the size of the cells progressively.

During this period (the S-phase of the life of the cell) desoxyribonucleic acid (DNA) is being synthesized, so that when the next cell division occurs there will be enough to enable the doubled chromosomes to separate into the two daughter cells. By the use of tritiated thymidine, which is incorporated into the DNA, followed by radioautography, an estimate can be obtained of the rate of cell proliferation in different tissues, both in the embryo and in the adult. This is possible because the amount of DNA per cell is constant, and an increase in the DNA content of a tissue or organ therefore indicates an increase in the number of cells it contains. However, it appears that cells can accumulate DNA at a very much slower rate than that found in tissues where active cell division is taking place, and it has been postulated that this slower rate may represent a turnover of DNA content quite apart from cell division.

Cells do not only divide and grow, they tend also to become 'differentiated'. In the early stages of the embryo there is no evidence of this: thus, if one of the daughter or granddaughter cells produced by division of the fertilized ovum is dissected out, it may itself prove capable of giving rise to a whole normal animal. Such cells are referred to as 'multipotent'. Later, however, their descendants lose this capacity, and become 'committed' to a certain specialized line of development by beginning to synthesize different proteins. Later still the majority of committed cells become further differentiated, so that their offspring ultimately give rise to specific tissues such as muscle, bone etc.

The tools by which we might recognize that differentiation has taken place are crude in the extreme, and we have to rely largely on visual appearances under the light or electron microscope. When the cell has the task of elaborating an easily recognizable product, for example, pigment, it is simple to detect the final stages of differentiation. But the precursors of such a cell may contain very little structural evidence relating to the future activities of their descendants, and attempts to detect the specialized apparatus with which we must assume they are equipped eventually peter out in failure as we trace the ancestral line further back towards the ovum.

In general terms undifferentiated cells tend to be uniform and nondescript in appearance, to be arranged in random patterns, to have relatively simple structure and functions, and to be highly adaptable under differing environmental conditions. In contrast, differentiated cells tend to be diverse and regular in appearance, to have a complex structure and specialized functions, and to be much more rigid and unadaptable. Some examples of differentiated cells which are referred to later are shown in PLATE 1, and it will be evident that their appearances are often strikingly dissimilar. Nevertheless, the one true criterion of differentiation is the functional one; a cell which retains only a single functional potentiality must be regarded as fully differentiated whatever its appearance may be.

Differentiation of a cell may entail differentiation of its responses to various growth-promoting stimuli. For example, the sex hormones secreted at the time of puberty have a much greater effect on the growth of certain tissues than on others, and the secondary sexual characteristics are produced by this selective stimulation [p. 104].

The factors which lead to a cell becoming differentiated are quite unknown, but they appear to include the influence of other cells in the body. Embryonic cells removed from the body of an experimental animal and cultured in appropriate media may remain embryonic in their characteristics and continue to grow and multiply without differentiating further, whereas if they had been left in their original site they would have differentiated.

We assume that the first stages of differentiation are chemical in nature, but very little indeed has been discovered about them, though sometimes chemical changes can be detected before structural changes become apparent, for example, in the developing heart of the chick. A differentiated cell usually acquires new functions, but it may at the same time lose old ones. Some of its genes are prevented from exerting their effects ('selective repression'), and others have their effects unblocked ('selective activation'). Differentiation and multiplication tend to be mutually exclusive, and the more specialized a cell is, the less likely it is to divide. An extreme example is the red blood corpuscle, which in the course of specialization to a respiratory function loses its nucleus, and with it any hope of division or indeed of repairing any damage it may suffer. Cells have therefore two pathways open to them. They can synthesize the enzymes appropriate to continuing in a 'mitotic cycle', or they can produce those appropriate to full functional activity and leave the mitotic cycle along a road which leads to ageing and death.

Tissues may be roughly classified into those in which there is a very active turnover of the cell population, those with a moderate turnover, and those with none. To the first group belong those tissues in which cells are constantly being lost or destroyed, and in which the total mass is kept constant; it includes epidermis and its derivatives, the lining of the gut, endometrium, transitional epithelium, the blood-forming tissues, and the cells producing the male and female sex cells. In such tissues the new cells are derived from special, relatively undifferentiated 'stem' cells which are collectively known as the growth fraction of the tissue.

In the second group are such organs as the liver, the kidney, and the exocrine and endocrine glands; in all of these there is no necessary loss of cells as a result of function, and their cell population thus does not constantly require to be restored. Growth occurs by division of the functioning cells themselves, and turnover is relatively slow. Nevertheless, these apparently specialized cells can, if appropriately stimulated, become temporarily able to divide and reproduce at a very remarkable rate, and this is particularly important in the repair of injuries [CHAP-TER 8]. The nature of the stimulus which causes such cells to divide is as yet unknown.

The third group of tissues and organs is small but important. Somatic muscle and nervous tissue differ from the organs in the second group in growing by cell division only in the early stages of their development; after a certain time cell division stops and further growth is produced merely by enlargement of the existing cells, which—at least theoretically—live as long as the body lives.

All division of cells other than the reduction division necessary for the maturation of the sex cells results from the process of mitosis, in which the chromosomes split longitudinally to preserve the diploid number in the daughter cells. This process is readily identifiable histologically [PLATE 2a], and the ratio of cells in mitosis to cells not in mitosis in a given tissue at a given time is known as its mitotic index. The mitotic index (and thus the rate of growth) is influenced by many factors. Thus it varies with age, sex, weight, diet, temperature, and time of day.

Growth of Intercellular Matrix

In the early embryo the intercellular matrix is scanty and amorphous. Later it increases in quantity, and comes to contain viscous substances which give it a gel-like consistency and enable it to hold a large amount of fluid. Chief among these substances are hyaluronic acid, a polysaccharide which retains water tenaciously and tends to become more viscous the more calcium is present in the fluid, and chondroitin sulphate, which is a firmer gel than hyaluronic acid and is particularly plentiful in cartilage.

Later still, as connective tissue cells become differentiated, the matrix is found to harbour fibres of two types. The more numerous of these are composed of collagen, a scleroprotein which by its ubiquity accounts for about one-third of all the protein in the body. Collagen fibres are laid down where firmness is needed— for example, in bone, skin, and tendon—and occur in bundles about 1–100 micrometres in diameter. Individual collagen fibres do not branch, but connexions exist between bundles because fibres tend to leave one bundle and join another: they take a straight or slightly wavy course through the matrix [PLATE 2b].

Collagen fibres are closely associated with cells known as fibroblasts, which appear to synthesize a predecessor protein called tropocollagen. This is extruded from the cell, and the microfibrils which go to form collagen are then assembled in the extracellular spaces. However, collagen fibres appear in the matrix of the embryo at a time when very few cells, let alone fibroblasts, are present. It is possible that sulphated polysaccharides may play an essential part in their formation, and other factors are known to be involved: for instance, vitamin C deficiency leads to defective collagen formation [pp. 151, 175].

The second type of fibre is the elastic fibre, about which even less is known. It is generally agreed that it is produced extracellularly, and it is possible that the presence of collagen in the neighbourhood is necessary for its formation. Elastic fibres are thinner than collagen fibres (0.2–1 micrometre in diameter), and branch freely; the individual fibres are relatively straight [PLATE 2b].

The formation of matrix and fibres continues throughout the whole period of growth, and is a vital factor in the repair of injuries [CHAPTER 8]. As growth proceeds, the matrix tends to lose some of its water content, and important changes occur in it as old age advances [CHAPTER 10].

Results of Cell Function

As cells differentiate they begin to manufacture their characteristic products or to carry out their specific chemical operations, and the results of their activity may contribute to the processes of growth. For example, the cells of the thyroid gland produce secretion which accumulates locally in the gland, causing it to enlarge by passive distension, and at the same time circulates throughout the body via the blood stream, with profound results on the growth of every tissue and organ.

PHASES OF GROWTH

There are four main phases in the growth of the body. In the early embryo everything is subordinated to growth, and there is little differentiation of function. This phase merges into a second one during which a kind of balance is struck between growth and differentiated functional activity; this phase lasts until maturity. After this a third phase supervenes, in which the main goal of the body is functional activity, and growth becomes a matter of making good losses occurring through wear and tear. The last phase occurs in old age, when growth becomes insufficient to keep the body in balance, so that cells are lost without replacement from various systems, and functions may become inefficient as a result [CHAPTER 10]. Finally, growth ceases with the death of the component tissues. In every phase the balance between growth and function can be upset by the advent of injury, which stimulates the amount of growth necessary to effect repairs.

Before birth, cell division is the main cause of growth, though other processes occur also. After birth the emphasis shifts to the enlargement of existing cells and the laying down of intercellular matrix. The extreme example of the contrast between antenatal and postnatal growth is afforded by the nervous system, in which it is thought that no new nerve cells appear after the sixth month of fetal life. If this is so, then we already have at birth all the nerve cells we can ever possess. In the postnatal phase of growth of the nervous system, the nerve cells add considerably to the amount of cytoplasm surrounding their nuclei, and use some of this additional material to increase the number and intricacy of their communications with other nerve cells, and to extend their peripheral processes to keep pace with the increasing distances these processes have to travel as the body enlarges.

The logistics of cell division during the four phases of growth are naturally complicated, and it will be clear that different proportions of stem cells and differentiated cells must be produced at different stages. If every stem cell divided into two more stem cells, as in the very earliest stages of embryonic development, the embryo would become a mass of unspecialized cells. If, on the other hand, every stem cell divided into two differentiated cells incapable of further division, growth would abruptly come to a halt. The complications of the systems of controls and checks which orderly growth requires can perhaps be illustrated by specific examples.

In the adult phase of growth in the epidermis of the skin, the cells which are lost from the surface by attrition are replaced by the activity of stem cells in the germinative layer in the depths of the epidermis [FIG. 55, p. 162]. Neglecting programmed cell death, these cells must produce 50 per cent of offspring which are differentiated and gradually proceed to the surface to be shed, and 50 per cent of offspring which remain as stem cells to continue the process of growth replacement. Any other proportion would mean that the skin would either show signs of overgrowth, or else fail to replace wear and tear and so become thinner. Of course this does not imply that every individual stem cell in the germinative layer must divide into one stem cell and one differentiated cell; all it means is that the over-all result of cell division in this situation must be equal numbers of stem and differentiated cells. What seems to happen in practice is that the growth fraction [p. 10] produces many more cells than are needed for the renewal of the tissue, and that the superfluous cells then undergo programmed destruction.

This is the situation in the adult phase of growth, but in the first two phases the number of stem cells in the germinative layer must increase, so that the growth of the skin keeps pace with that of the body which it covers. At the same time the skin grows in thickness, so that more differentiated cells must also be produced by the activity of yet more stem cells. Further, the skin becomes thicker in some places than in others, so that local as well as general controls are necessary. In contrast, the thickness of the skin diminishes in old age because the stem cells are unable any longer to keep pace with the surface losses. In all these phases the fine adjustment maintaining the balance between division and differentiation is a razor edge between 'normal' and 'abnormal'. If the adjustment becomes disorganized, then uncontrolled local or general growth may take over, or further growth may cease [CHAPTER 9].

There is now evidence that the rate of cell division in the germinative layer is controlled by a chemical feedback system. It is possible to extract from epidermis a substance which delays DNA synthesis and suppresses mitosis; it is supposed that the accumulation of this material, known as a chalone, serves to maintain the balance between cell division and differentiation. There is every likelihood that growth and replacement in other tissues and organs, such as the wall of the gut, the kidneys, and the lungs, depend on similar controlling chalones formed by the actively dividing cells. It has not yet been conclusively proved that each of these chalone systems is specific for its own particular tissue, but this is probably the case. At all events the normal balance between mitosis and function in every tissue is accurately controlled, and on this view of the hormonal control of such cycles of growth and death as the menstrual cycle [p. 167] is only a more generalized extension of the universal local chemical control systems.

Another example of a delicate control mechanism is afforded by the ability of the body to maintain approximately constant the number of red blood corpuscles in the blood stream. Should the oxygen tension in the blood be reduced, perhaps by the subject going to live at a high altitude, the blood-forming stem cells in the marrow of the bones are stimulated to divide and produce more red cells, the number being apparently adjusted precisely to the respiratory needs. A similar increase in stem cell division in the marrow follows loss of blood by haemorrhage. We can say that the blood-forming cells respond to oxygen lack or to some chemical stimulus resulting from oxygen lack, but we have no idea of the basic processes within the stem cell which cause it to divide, or of the antagonistic processes which damp down its activity and prevent it from manufacturing so many red cells that the viscosity of the blood would rise dangerously.

Enough has perhaps been said to indicate that cell division is subject to a whole series of controls at each stage of development, that some of these controls are highly specialized, and that most of them are as yet mysterious.

GROWTH CURVES

The total amount of growth achieved obviously depends on the time for which growth proceeds and also on the speed of growth per unit time. Among the primates the duration of postnatal growth seems to be correlated with the age at which sexual maturity is reached; the greater this age, the longer the period of growth. Growth in the lemurs lasts for two or three years, in the monkeys for seven or eight years, in the great apes for ten or eleven years, and in man for up to twenty or more years.

Measurements taken on a single individual at intervals can be plotted against time to produce a graph of progress [FIG. 4], whether they are derived from the whole body or from one of its component parts. A graph of this sort, in which the actual measurement at a given time is plotted against the time at which it was taken, is sometimes called a 'distance curve', since any point on it indicates the distance the body has travelled along the road to maturity. It is clear that the curve must flatten out to a plateau as growth ceases.

Another way of presenting the same data is shown in FIGURE 5, in which the increments of growth (that is, the amounts added in specific time intervals) are plotted against time. Such a curve shows the variation in the rate of growth with time, and is therefore known as a 'velocity curve'. Ultimately, such a curve must tend to zero as growth ceases.

If a growth curve is derived from a single individual, or from repeated measurements on the same group of individuals, it is said to be based on 'longitudinal' data. Such information is naturally difficult to get, since every subject must report at regular intervals to be measured, and in consequence gaps in the records are only too common. It also requires a great deal of patience, since results cannot be evaluated for several years.

It is much easier to obtain what are called 'horizontal' or

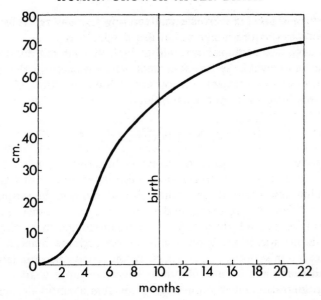

FIG. 4. Distance curve for stature before and after birth. Each point on the graph represents the mean height of several embryos or children of that particular age: the curve is therefore said to be derived from cross-sectional data (see text). (From Thompson, D'Arcy W. (1942) *Growth and Form*, 2nd edn, London, Cambridge University Press, by kind permission of the publishers.)

'cross-sectional' data. Measurements are made of several children in each age group, and these are then combined to form a cross-sectional picture of the various age-groups in the community at the time of the investigation. In such work the sample can be accurately controlled; since each child is measured only once there is no difficulty about children not returning to be measured, and the results are immediately available. Nevertheless, they are not so useful from a statistical point of view as are longitudinal data, and may actually be misleading if one wishes to investigate such matters as changes in development or alterations in velocity [FIG. 13, p. 28]. For such work, longitudinal studies are essential.

'Mixed longitudinal' investigations are basically longitudinal ones in which some of the children dropped out for one reason or another, and had to be replaced by others. Instead of having to discard all information obtained from those who did not complete the whole course, the investigator can now use special statistical methods to obtain a meaningful result.

A curve such as that shown in FIGURE 4 is an obvious challenge to the mathematically minded, and many attempts have been made to express growth curves in mathematical form. A very brief discussion of these must suffice to indicate the sort of problem involved.

The initial stages of growth after fertilization involve cell division only, and proceed in geometrical progression, one cell giving rise to two, two to four, four to eight, and so on. We have already noted that growth gradually slows down towards a halt as the adult condition is approached. It follows that the curve of growth from the ovum to the adult must be S-shaped in its general outline. We are here concerned particularly with the part of it which describes growth after birth, and this results, in varying combination, from cell division, cell enlargement, and production of matrix. The first of these factors suggests geometrical progression, and the other two suggest arithmetical progression: it will already be clear that no simple equation is likely to help us very much.

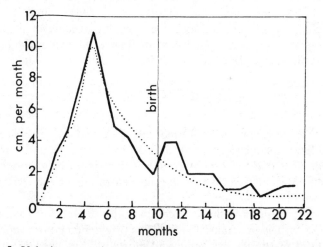

FIG. 5. Velocity curve for stature before and after birth: derived from the same data as FIGURE 4. Each point on the graph represents the mean increase in length during that particular month. (From Thompson, D'Arcy W. (1942) *Growth and Form*, 2nd edn, London, Cambridge University Press, by kind permission of the publishers.)

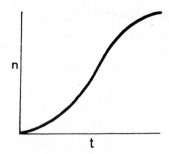

Fig. 6. Curve of logistic growth
n = number of cells present
t = time

If cells are grown in tissue culture, the initial geometrical progression does not continue indefinitely, for the supply of foodstuffs is not unlimited, and the environment is progressively polluted by the waste products of the cells. There is thus a gradual slowing down of multiplication, until finally it stops. If there is no renewal of the environment, the cells resulting from the process of multiplication must eventually die, so that their number gradually returns to zero. If, however, the environment is continuously renewed, the population of cells ultimately approaches the maximum which the available volume of medium will support, so producing an S-shaped curve known as the curve of logistic growth [Fig. 6], which is described by the equation:

$$W = \frac{a}{1 + be^{-kt}}$$

where W is weight and t is time, and a, b, and k are constants. This curve is sometimes fitted to human growth data, and expresses the idea that the rate of growth is dependent on the number of cells present and also on the concentration of the available food supply.

Another type of S-shaped curve was formulated by Gompertz in 1825 for use in actuarial tables. The original equation, which is still used in studies of ageing and mortality, has been modified many times for application to data on growth. One form may be written:

$$W = ae^{-be^{-kt}}$$

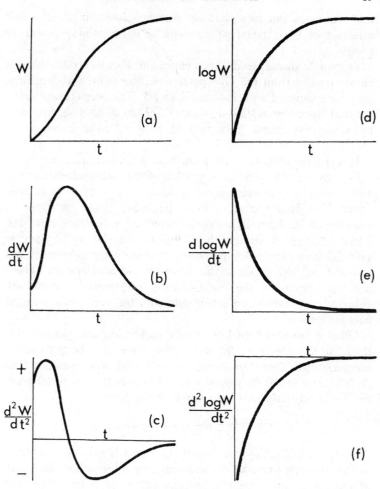

FIG. 7. A Gompertz function. (*a*) Distance curve; (*b*) Velocity curve; (*c*) Acceleration curve; (*d*) Curve of specific growth; (*e*) Velocity curve of specific growth; (*f*) Acceleration curve of specific growth. The scales of the ordinates have been adjusted to make the height of all the graphs the same.

$$W = \text{weight}$$
$$t = \text{time}$$

(From Medawar, P. B. (1945) Size, shape and age, in *Essays on Growth and Form*, Oxford, Clarendon Press, by kind permission of the author and publishers.)

and expresses the idea that an organism loses, in equal small intervals of time, equal proportions of its remaining power to grow.

FIGURE 7a shows the distance curve and FIGURE 7b the velocity curve derived from the Gompertz equation written above; they may be compared with FIGURES 4 and 5. The acceleration curve derived from this equation [FIGURE 7c] has a positive, followed by a negative phase, returning to zero with the cessation of growth.

In experimental work on growth it is sometimes mathematically convenient to plot the logarithm of the measurement against time, thereby producing what is known as a 'specific growth' curve. This is a record of how the living tissue measured is multiplying itself; if it multiplies itself at a constant rate, the curve becomes a straight line. FIGURE 7d shows the specific growth curve of the Gompertz equation we are considering, and FIGURES 7e and 7f show the velocity and acceleration curves derived from it. If the multiplication rate were constant, the velocity curve would be a constant, and the acceleration would become zero.

Other equations with less readily understandable implications than the Gompertz or logistic curves have also been fitted to human growth data. For example, it is found experimentally that growth from about the age of 1 year to about the age of 10 years can be fitted fairly well by a curve of the form:

$$W = a + bt + c \log t$$

though the biological meaning of the terms is far from clear.

The practical value of the mathematical approach to the study of human growth would be greatly enhanced if from the best fitting equations there could be derived a clear indication of the sort of biological processes which must underlie the measurements. So far this has not been possible.

2

Growth in Height and Weight

'With sobs and tears he sorted out
Those of the largest size'

LEWIS CARROLL

GROWTH IN HEIGHT

Growth Curves and Growth Charts

The human ovum measures about 100 micrometres in diameter, and is just visible to the naked eye. At birth, a baby is about 50 cm (20 in) long—roughly five thousand times as long as the ovum. Further growth in height after birth may take the adult to a height of (say) 175 cm (5 ft 9in)—three and a half times the length of the baby.

Growth in height, like all other human measurements, is not uniform throughout life. The maximum rate of growth occurs before birth [FIG. 5], in the fourth month of fetal life, when it is about 1·5 mm a day; thereafter there is a progressive slowing down, though at birth the baby is still growing very fast indeed compared with the infant and child.

FIGURE 8 shows curves derived from the data obtained by Count Philibert de Montbeillard, who measured the height of his son from 1759 until 1777. These data were subsequently published by Buffon, and are the earliest longitudinal measurements we possess. They show the progress made by an eighteenth-century French boy, raised in the country, and probably comfortably off; it is interesting to compare them with the average progress of modern children in Britain [FIG. 11].

A number of difficulties encountered in the measurement of height in children and adults should perhaps be mentioned at this

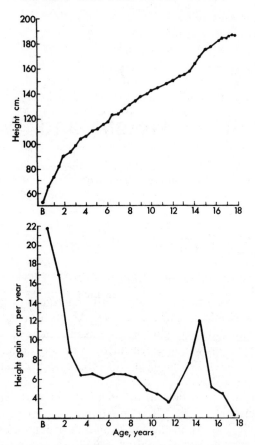

Fig. 8. Growth in height. Progress of the son of Count Philibert de Montbeillard from birth to the age of 18 years. Upper graph, distance curve showing height reached at each age: lower graph, velocity curve, showing annual increments in height. (From Tanner, J. M. (1962) *Growth at Adolescence*, 2nd edn, Oxford, Blackwell Scientific Publications Ltd., by kind permission of the author and publishers.)

point. Errors will be introduced if the apparatus used is not rigid. The subject must stand on a firm horizontal floor against a solid vertical wall and be measured with the aid of a flat horizontal surface rather than a thin bar. His posture must be standard, with the lower border of the orbits level with the external auditory meati. The examiner should press upwards on his mastoid processes and tell him to make himself as tall as possible. Height depends to a considerable extent on the width of the intervertebral discs of the vertebral column [p. 79], and these become compressed by the gravitational strain imposed by the upper part of the body as the day goes on, with the result that, unless the subject spends the day in bed, he measures less in the evening than he does in the morning. This diurnal variation may be as much as 2 cm, even in children. It is thus essential that all measurements of height should be made at the same time of day. Another complication is that height varies according to the tension of the muscles which control the posture of the pelvis and vertebral column; if they are weakened, perhaps through illness, the height of the individual may alter. A mild scoliosis (p. 136), which may not be clinically obvious, can be a source of difficulty. Finally, it is most desirable that the same person should make every measurement, so that systematic errors of technique or interpretation may be minimized. This is not always easy when dealing with large numbers. Under the best circumstances, with no complicating factors, the error in measuring an individual is anything up to 3 mm.

Given adequate data from a sufficiently large group of children, usually at least 1000, it is possible to construct a chart which shows the relationship of height to age in the group studied [FIGS. 9 and 10]. Because of individual variation, there is a considerable spread in the figures for a given age, and this is indicated by the centile lines on the charts. These show the percentages of children whose heights lay on or below the labelled line. Thus, 25 per cent of the group had heights which lay on or below the labelled 25th centile line. Such charts are of course cross-sectional [p. 15] in nature, and it must be stressed that the centile lines do not correspond with, or even lie parallel to, the distance growth curves of any particular child. Longitudinal measurements on an individual [FIG. 11] yield growth curves of substantially different shape.

Charts of height for age are invaluable in paediatrics, but their use requires a good deal of caution. If a measurement shows that a child lies outside the 97 per cent centile—i.e. that he is taller than 97 per cent of the children in the 'standard' group—there is a temptation to suppose that his progress has been 'abnormal', and that there must be something wrong with him. Now in the

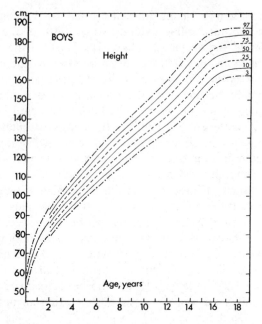

F IG. 9. Cross-sectional standards for height attained at each age (English boys). The central line represents the mean, or 50th centile. The two dashed lines above and below it represent the 75th and 25th centiles respectively—i.e. 25 per cent of the sample fell below the lower line and 25 per cent above the upper line. Other centile lines are also provided. (Modified from Tanner, J. M., Whitehouse, R. H. and Takaishi, M. (1966) *Arch. Dis. Childh.*, *41*, 454–71, by kind permission of the authors and the editor.)

first place we must be sure that we are comparing like with like when we use the chart, for children in different circumstances do not necessarily grow in the same way [p. 155], and a chart derived from American children may look very different from a chart representing the growth of Indian children. But even when the appropriate chart is used, we can only say that a child outside the 97 per cent line is 'unusual'—after all, one child in every 33 of the 'standard' group also fell outside this parameter. The evidence of the standard chart is only one piece among many: for example, the family history is equally important, for there are many genetic deviations among growth patterns.

In the first year after birth body length increases by about 50 per cent, to 75 cm (about 30 in), and in the second year another 12–13 cm (about 5 in) or so are added. Thereafter growth in height settles down to a rate of about 5–6 cm every year. For comparative purposes a figure known as the relative growth rate is sometimes used. The 'absolute' rate is the amount of growth in a given period divided by the length of the period. The 'relative' rate is obtained by dividing the absolute rate by the initial height and expressing the result as a percentage. Thus a child 120 cm tall who grew 9 cm in a year and a half would have an absolute growth rate of 6 cm per year, and a relative rate of 5 per cent per year.

Round about the time of adolescence the fairly steady rate of increase in height rises suddenly and markedly. This 'adolescent spurt' in height is not easily discernible in cross sectional distance curves of growth, but becomes much more immediately evident in individual longitudinal curves [FIG. 11], and is shown very

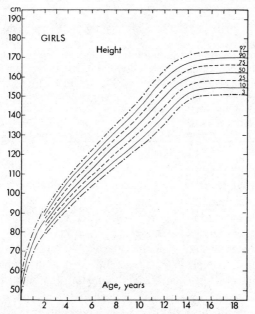

FIG. 10. Cross-sectional standards for height attained at each age (English girls). Centiles as in FIGURE 9. (Modified from Tanner, J. M., Whitehouse, R. H., and Takaishi M. (1966) *Arch. Dis. Childh.*, *41*, 454–71, by kind permission of the authors and the editor.)

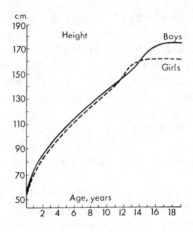

FIG. 11. Typical individual height-attained curves for English boys and girls. (From Tanner, J. M., Whitehouse, R. H., and Takaishi, M. (1966), *Arch. Dis. Childh.*, *41*, 454–71, by kind permission of the authors and the editor.)

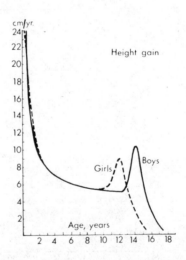

FIG. 12. Typical individual velocity curves for height: English boys and girls. (From Tanner, J. M., Whitehouse, R. H., and Takaishi, M. (1966) *Arch. Dis. Childh.*, *41*, 454–71, by kind permission of the authors and the editor.)

strikingly in longitudinal velocity curves [FIG. 12]. The spurt begins about the age of $10\frac{1}{2}$ to 11 years in girls, and $12\frac{1}{2}$ to 13 years in boys, though wide variations are possible [FIG. 13]; for example, the timing of the onset of the male spurt ranges from $10\frac{1}{2}$ years to 16 years. In both sexes it lasts for 2 to $2\frac{1}{2}$ years. During the spurt boys add something like 20 cm (about 8 in) to their height, mostly because of growth of the trunk [p. 123]. At the peak of growth velocity, round about 14 years of age, they may be growing at the rate of 10 cm a year. Girls gain a matter of 16 cm during the spurt, with a peak velocity about 12 years of age.

The conclusion of the spurt is followed by a rapid slowing of growth; girls reach 98 per cent of their final height by the average age of $16\frac{1}{2}$, whereas boys do not reach the same stage until the age of about $17\frac{3}{4}$: again there is wide variation around the means. Up to the time of the adolescent spurt there is little difference between the average heights of boys and girls. Because the spurt begins earlier in females there is an age at which girls become taller (and heavier) than boys of the same age; in Britain today this age is about 11 years. The balance is redressed by the age of 14, when boys overtake girls in height, though they do not become heavier until some time later.

It is important to realize that if measurements taken from several children are combined on the same graph, the adolescent spurt is smoothed out and becomes less dramatic [FIG. 13]. This is because the spurt occurs at different times in different children, and a much better picture of what is going on is obtained from the study of individuals. To overcome this difficulty, another way of presenting the information is sometimes used. The peak velocity in each child, irrespective of age, is superimposed on the peaks taken from the others, and this is used as zero time, measurements backwards towards birth being recorded as -1, -2, -3 years, etc., and measurements forwards towards maturity being recorded as $+1$, $+2$, $+3$ years, etc. [FIG. 13].

Man is apparently alone in having such a long quiescent interval before the adolescent spurt, and the reason for this has been the subject of much discussion. It has been tentatively suggested that the main reason for the difference is a need to postpone puberty in order to allow time for the necessary maturation of the more complex human brain. In other words, the

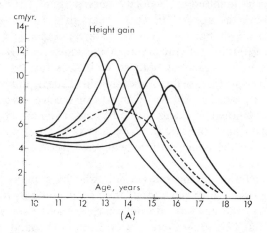

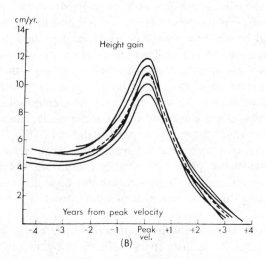

Fig. 13. Adolescent spurts in height. (A) individual velocity curves of five boys. The dashed curve is obtained by averaging the individual values at each age. Notice how this average curve fails to show the dramatic nature of the spurt in the individual case. (B) The same curves superimposed so that their peak velocities coincide (see text). (From Tanner, J. M., Whitehouse, R. H., and Takaishi, M. (1966) *Arch. Dis. Childh.*, *41*, 454–71, by kind permission of the authors and the editor.)

delay may be correlated with the fact that human survival depends on learning, which takes time, rather than on instinct, which does not.

Some children exhibit another transitory upheaval of the steady childhood growth in height round about the age of 6 or 7 years; this is the so-called 'mid-growth' spurt. Very little is known about this, and it does not appear to be a feature of every growth curve; it is much less impressive than the adolescent spurt.

Although noticeable growth in height stops round about 18 in the female and 20 in the male, there is evidence to show that a certain amount of growth continues slowly for many years after this. The vertebral column is said to continue to grow up to about 30 years of age, but the total increase in height due to this cause is little more than 3–4 mm (about $\frac{1}{8}$ in), and the difficulties of assessing measurements of so small an amount will at once be apparent. It is also said that most head measurements continue to increase very slowly up to the age of 60, but again the increase from the age of 20 is only about 2–4 per cent of the 20-year values. These very minor increases, if they really exist, are negligible in comparison with the loss of height which becomes apparent for the first time shortly after middle age, and is due to the degeneration of the intervertebral discs and probably also to the loss of thickness of the joint cartilages of the lower limbs [p. 225].

It is not yet known precisely what stops growth in height: it is usually said to be due to the influence of the hormones derived from the gonads [p. 147], but a great deal remains to be discovered about this. Certain cold-blooded animals, such as fish, appear never to stop growing, though they do so at a considerably reduced rate after attaining maturity.

Prediction of Adult Height

By analysis of data derived from a sufficient number of children measured longitudinally it becomes possible to say what percentage of their final adult height has been achieved at any given age, and thus to produce a table which will predict, at least roughly, the future height of an individual from measurements taken in childhood [TABLE 1]. The predictive value of such a table is nil at birth, for the birth length, like the birth weight, is

TABLE 1

PERCENTAGE OF MATURE HEIGHT ATTAINED AT DIFFERENT AGES

CHRONOLOGICAL AGE (YEARS)	PERCENTAGE OF EVENTUAL HEIGHT	
	BOYS	GIRLS
1	42·2	44·7
2	49·5	52·8
3	53·8	57·0
4	58·0	61·8
5	61·8	66·2
6	65·2	70·3
7	69·0	74·0
8	72·0	77·5
9	75·0	80·7
10	78·0	84·4
11	81·1	88·4
12	84·2	92·9
13	87·3	96·5
14	91·5	98·3
15	96·1	99·1
16	98·3	99·6
17	99·3	100·0
18	99·8	100·0

The percentages given are derived from a longitudinal study of 150 boys and girls carried out by the University of California (Bayley, N. (1956) Growth curves of height and weight for boys and girls, scaled according to physical maturity, *J. Pediat., 48*, 187–94.)

(Reproduced by kind permission of the author and publishers.)

considerably influenced by the environment of the fetus in the womb. For example, a baby ultimately destined by its genetic make-up to be tall may measure relatively little at birth if it is premature, one of a multiple birth, or born to a young mother [cf. p. 32].

But by the second birthday the child has joined the genetic curve which basically determines his height [p. 140], and predictions become possible. A single measurement is of little use, for the rate of growth in an individual varies from time to time. A child who is on the 50th centile at the time of measurement may be just starting a period of intensive growth or may be about to slow down. He has to be observed for a year or more, because of seasonal variations in growth [p. 156], and if several readings of height are taken in this period his rate of growth can be compared with standard centile charts of growth velocity. These are sometimes confusing, if the timing of his adolescent spurt does not conform to the mean [FIG. 13].

Predictions are improved if the skeletal maturity [p. 98] of the child is known. In 1975 Tanner and his collaborators calculated for British children a series of equations which utilized the height, the chronological age, the skeletal age, and the onset of menstruation [p. 106] to predict ultimate height. Because of the strong genetic influence [p. 140] on height the height of the parents was also taken into account. Even this elaborate system could only predict adult height in boys aged 4 to 12 years to within ±7 cm in 95 per cent of cases. The accuracy improved to ±6 cm at the age of 14. Since the 95 per cent range of adult male height in Britain is ±12·5 cm, such predictions are considerably better than nothing, but they are still disappointingly vague. The figures for girls are similar except that prediction becomes more accurate earlier; the eventual height of girls of 12 or 13 who have begun menstruating can be predicted within ±4 cm and ±3 cm respectively.

The importance of such measurements is more than academic. For example, ballerinas should be less than 167 cm (5 ft 6 in) tall, with a permissible maximum of 175 cm (5 ft 9 in) and a minimum of 157 cm (5 ft 2 in). Since attendance at ballet school begins in the early years of childhood, it is obvious that a great deal of bitter disappointment might be avoided if a reliable prediction showed that the likely adult height of the candidate was outside the acceptable limits.

Height restrictions obtain in other occupations. Chorus girls are rarely accepted if they are less than 163 cm (5 ft 4 in) tall, and nurses are often excluded if they are less than 152 cm (5 ft) tall, for the reason that people of this size are not much use at lifting heavy patients. A policeman must be more than 173 cm (5 ft 8 in), and a policewoman more than 160 cm (5 ft 3 in), while an Army parachutist must be less than 190 cm (6 ft 3 in). Studies of this kind have therefore a certain limited value in the selection of careers.

Apart from such considerations, excessively tall girls and excessively small boys suffer considerable psychological distress, and if the probable height is outside acceptable limits an attempt can be made to modify their future growth before it is too late [p. 148].

GROWTH IN WEIGHT

Growth Curves and Growth Charts

At birth, after 280 days of gestation, a baby weighs on the average about 3·4 kg (7½ lb), with a range of about 2·7–4·5 kg (roughly 6–10 lb). This is something like three thousand million times the weight of the ovum, and gives an idea of the furious growth activity before birth. The maximum growth rate, unlike the maximum rate of growth in height, is not achieved until shortly *after* birth [Fig. 15], but it very soon slackens off markedly, and in the twenty years or so of growth from birth to maturity the birth weight increases by a further factor of only about 20 times—say to an adult weight of 68 kg (150 lb).

The weight at birth is more variable than the length, and reflects the maternal environment more than the heredity of the child. Like body length, it is less when the mother is young and when the baby is premature. Full term females are on the average about 140 g (roughly 5 oz) lighter than full term males, and a twin weighs on the average about 680 g (roughly 1½ lb) less than a singleton; a triplet is about 340 g (¾ lb) lighter than a twin. Small mothers tend to have small babies, irrespective of the size of the

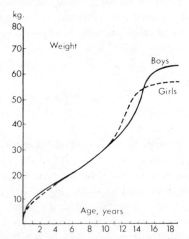

Fig. 14. Typical individual weight-attained curves for English boys and girls. (From Tanner, J. M., Whitehouse, R. H., and Takaishi, M. (1966) *Arch. Dis. Childh.*, *41*, 454–71, by kind permission of the authors and the editor.)

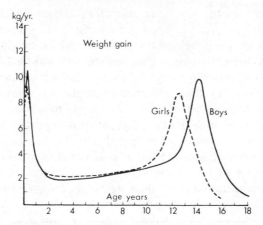

Fig. 15. Typical individual velocity curves for weight: English boys and girls. (From Tanner, J. M., Whitehouse, R. H., and Takaishi, M. (1966) *Arch. Dis. Childh.*, **41**, 454–71, by kind permission of the authors and the editor.)

father, and mothers in a low socio-economic group have smaller babies than those with a higher rating. The rank of the child in the family is a factor in its birth weight, later children being somewhat heavier than the first-born. The birth weight is independent of the mother's diet unless there has been severe malnutrition, since the fetus will draw on the mother's tissues for its needs. Women who smoke during pregnancy have smaller babies than those who do not, by about 170 g (6 oz) on the average.

Immediately after birth the diminished intake of fluid leads to a transient loss of about 5 per cent of the birth weight; this loss is usually made good in about 10 days, and has no connexion with the ability of the body to grow.

It seems that British babies are now putting on weight in the first few months more rapidly than they used to. In London in 1973 more than 40 per cent of healthy babies 3 months old weighed over 7 kg (15½ lb), whereas twenty years earlier fewer than 10 per cent weighed as much at this age.

By the end of the first year the birth weight has approximately tripled, and by the end of the second it has quadrupled. After this it settles down, like the growth in height, to a relatively steady annual increase, in this case of about 2·25–2·75 kg (5–6 lb) a year, until the onset of the adolescent spurt [Figs. 14 and 15].

During the spurt, boys may add 20 kg (about 45 lb) to their weight, and girls 16 kg (about 35 lb); the peak velocity for the spurt in weight lags behind the peak velocity for height by about 3 months: the child first begins to shoot up, and only later does he start to fill out. Similarly, body weight does not reach its adult value until some time after adult height has been attained. Weight increase is usually taken as a good indicator of satisfactory progress of growth in a child, but it is important to distinguish between increases in weight due to healthy deposition of bone and muscle and those due to unhealthy laying down of fat.

Measurements on a large group of children can be combined into a cross-sectional chart showing weight for age in exactly the

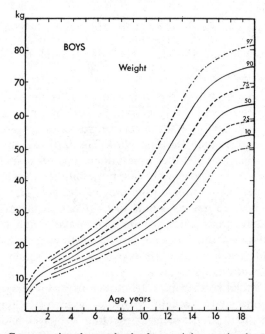

FIG. 16. Cross-sectional standards for weight attained at each age: (English boys). The central line represents the mean, or 50th centile. The two dashed lines above and below it represent the 75th and 25th centiles respectively; i.e. 25 per cent of the sample fell below the lower line and 25 per cent above the upper line. Other centile lines are also provided. (Modified from Tanner, J. M., Whitehouse, R. H., and Takaishi, M. (1966) *Arch. Dis. Childh.*, *41*, 454–71, by kind permission of the authors and the editor.)

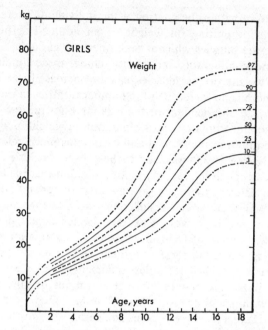

FIG. 17. Cross-sectional standards for weight attained at each age: (English girls). Centiles as in Figure 16. (Modified from Tanner, J. M., Whitehouse, R. H., and Takaishi, M. (1966) *Arch. Dis. Childh.*, *41*, 454–71, by kind permission of the authors and the editor.)

same way as such charts are constructed for height [p. 23]. Weight for age charts [FIGS. 16 and 17] are put to the same sort of use as height for age charts, and are subject to the same limitations.

The 'ponderal index' is a figure sometimes used to relate the weight and height of a growing child. It is defined as one hundred times the cube root of the body weight divided by the stature. This index is highest in the first two years of life, and thereafter gradually falls until towards the end of the adolescent spurt, when it rises again to about the same level as it had at the age of 8 or 9 years.

Growth in Weight after Maturity

Like height, weight may increase after maturity has been reached, but the results are much more dramatic, and are to a certain extent within the control of the individual. There are two main ways of putting on weight as an adult. The first is by exercising the muscles while at the same time eating a diet which is satisfactory in quantity and quality. Under these conditions the size and weight of the muscles can be increased. So far as is known, there is no increase in the number of fibres in the muscle, but simply an enlargement of the existing ones [p. 198].

It is also, unfortunately, possible to put on weight of a different kind by eating too much. If the diet contains more calories than the body can expend, specially if these excess calories are supplied by carbohydrate, fat may be deposited in the tissues. At the time of middle age there is a slowing down of general metabolism [p. 114], but there may be no comparable change in appetite. There may also be a lack of exercise, which combines with the unnecessary intake of food to cause the accumulation of excess fat which is one of the biggest problems of so-called civilization.

One cannot deposit fat unless sufficient calories are taken in the diet, but many people who eat enormous quantities remain enviably thin, and much work has still to be done on the causes of obesity. There are many reasons for eating too much food, and simple greed is only one. Another is anxiety; people who worry too much may calm themselves by overeating. An unusual reason is the professional requirements of Sumo wrestling in Japan. The optimum weight for a Sumo wrestler is about 140 kg (over 300 lb), and men who are in training to become wrestlers eat more than 5000 Calories daily. They are said to die ten years younger than the average Japanese, and are prone to diseases of the liver and kidneys as well as to diabetes. Competitors in the 'heavy' events at the Olympic Games also find it necessary to put on weight, and may also be at risk for similar reasons. Such athletes sometimes use hormones to assist the deposition of protein. Formerly androgens [p. 146] derived from the gonads or suprarenal cortex were used, but nowadays synthetic 'anabolic' hormones are in favour.

Another cause of increase in weight in adult life is pregnancy. A normal young woman having her first baby in Britain may put on about 13–14 kg (about 28 lb) on the average; there is a very wide range of variation. Four or five kg (about 10 lb) of this is due to the infant, the surrounding fluid, and the placenta, and the greater part of the remainder is accounted for by the growth of the uterus and the breasts. After the birth, and after the uterus has reverted to its resting state, her weight will still be about 5 kg (11 lb) more than it was originally; this is due to the increase in the size of the breasts, and to the stores (mostly of protein and fat) laid down for the feeding of the infant. Some of this weight increase is usually permanent, and the amount of the permanent deposit often increases with each successive child. Married women with children thus grow in weight more than their child-less sisters.

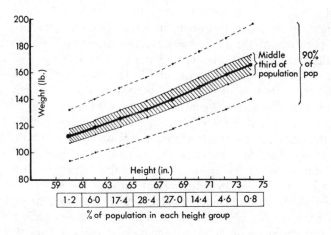

Fig. 18. Relationship of adult height to weight. Data from Scottish men aged 18 to 40 in 1954. (From Sinclair, D. (1976) Growth, in *Textbook of Human Anatomy*, 2nd edn, ed. Hamilton, W. J., London, Macmillan and Co. Ltd., by kind permission of the publishers.)

The variable amounts of muscle development and fat deposition are two of the main factors which make the relationship between height and weight in the growing child a somewhat elastic one. Nevertheless, in the adult there is a fairly good correlation between height and weight, and the results of a Scottish survey are shown in Figure 18.

It is important, when consulting tables of weights for heights, to distinguish between *average* and *ideal* figures. The average figures, both in the United States and in Britain, are steadily drawing away from the ideal figures laid down by such bodies as the Metropolitan Life Insurance Company on the basis of maximum longevity. The average weight of a representative American male of medium build aged between 25 and 29 years is now about 4·5 kg (10 lb) greater than the insurance company's ideal.

CHANGES IN BODY COMPOSITION

It is not surprising that there should be variation in measurements of weight. In regard to height, there are only three components to be considered [p. 57]—the bones, the cartilages, and a limited amount of connective tissue and skin. In the case of weight, every tissue and organ in the body is involved. Roughly 50 per cent of the adult body weight is due to water located inside the cells, and a further 15 per cent to water in the tissue spaces, making a total of about 65 per cent. In the new-born baby 80 per cent of the body weight is water, 35 per cent intracellular and 45 per cent extracellular. Measurement of body weight at a given moment therefore depends quite considerably on the degree of hydration of the tissues, and this in turn depends on many different influences. For example, it is reported that a woman taking contraceptive pills containing oestrogens may gain up to 13 kg (about 28 lb) in weight through the water-retaining effect of the oestrogens. Conversely, as every boxer and jockey knows, a considerable amount of weight may temporarily be lost by violent sweating, as in severe exercise or in a Turkish bath.

TABLE 2

WEIGHT DISTRIBUTION IN THE ADULT

ORGANS OR TISSUES	PERCENTAGE OF BODY WEIGHT
Muscles	43
Fat	14
Bone and marrow	14
Viscera	12
Connective tissue and skin	9
Blood	8
	100

The figures are averages relating to a male weighing approximately 70 kg (154 lb)

Patients can also lose water from the body by taking diuretics or purgatives. Other variables in the measurement of weight include the intake of meals, the amount of urine in the bladder, and the amount of faeces in the intestine.

TABLE 2 shows the percentage composition by weight of a 'standard' man weighing 70 kg (154 lb). The figures are derived from the recommendations of a Commission on Radiological Protection, and, like all biological averages, are meaningless when applied to an individual. However, they do give an idea of the contribution made by different tissues and organs. But not only do the absolute amounts of these tissues increase with age, their relative proportions vary at different ages. Bone weight remains fairly constant in relation to total weight, but in the infant the muscles account for only about 25 per cent of total weight, whereas in the adult the figure is over 40 per cent. On the other hand, the viscera are relatively heavy in the infant. A good deal of effort has recently been put into trying to find out the composition of the body at different ages, and this has been investigated by several methods. The most obvious is by direct weighing of viscera, muscles, etc. This can be done in dead bodies, but the results are much more difficult to interpret than might be imagined, partly because the cause of death may itself produce changes in the composition of the body.

There are similar difficulties in the chemical analysis of tissues. For example, it is possible to analyse the composition of an amputated limb, but this result cannot be applied with confidence to an estimate of the composition of the whole body. In a rather different category are the results of chemical determinations of the blood volume, which can be made in intact healthy children, and are therefore probably more accurate as an indication of the normal situation.

Radiology has been useful in the estimation of the proportions of bone, fat, and muscle in the limbs. A soft-tissue radiograph will show the relative widths of the shadows cast by the bone and the muscle, and the amount of fat can be obtained from the distance between the outer border of the muscle shadow and the skin. It is found that there is good correlation between measurements of fat made in this way at different sites in the body, but the correlation is less satisfactory for muscle. Formulae exist by which measurements made at selected places can be transformed into estimates of the proportions of these tissues in the whole body.

Another way of determining the amount of fat in the body is to measure with special calipers the width of skin folds pinched up at different places. Two sites frequently used are on the back of the arm over the triceps muscle and on the back of the trunk below the angle of the scapula. Estimates of total body fat based on such measurements, like those derived from radiographic measurements, are little more than informed guesses, since there is no stable and reliable correlation between subcutaneous fat and the amount of fat in the abdominal cavity. But both methods do yield a satisfactory estimate of subcutaneous fat, and this can be used to make comparisons between different individuals and different age groups.

Despite the above remarks, the thickness of the triceps skinfold correlates well with body density (obtained by an underwater weighing technique), and is now used as an index of total body fat. Other ways of determining this quantity are still less direct and of uncertain accuracy. For example, the total body water can be measured and from this the 'lean body mass' can be calculated. This is then subtracted from the total body mass to give a figure representing the total body fat.

Finally, it is possible to calculate the total amount of muscle in the body very roughly by measuring the power of a limited group of muscles such as those of the upper limb. Similar caution must be observed in relating such measurements to the total musculature, but again direct comparisons between individuals can be made without objection.

THE MERITS OF SIZE

The size range of man is no accident. The size of mammals is physiologically determined by their mode of respiration, and man cannot be as small as an insect, just as an insect cannot be as large as a man. Nor can a man be as large as an elephant. It was Galileo who pointed out that 'Nature cannot construct an animal beyond a certain size, while retaining the proportions and employing the materials which suffice in the case of a smaller structure.' The reason for this is to be found in the same laws of mathematics and physics that limit the size of cells. The volume of similar bodies, and hence their weight, increases with the cube of the linear dimensions: a trout 12·5 cm (5 in) long is nearly twice as heavy as one 10 cm (4 in) long. At the same time, the cross-sectional area of the bones which support this weight only increases with the square of the linear dimensions.

J. B. S. Haldane once calculated that the giants in the illustrations in his copy of the *Pilgrim's Progress* must have been about 60 feet tall. Now these giants would have a weight of about 1000 times that of a normal man, but the cross-section of their bones would be only 100 times greater. Hence every square centimetre of bone in the giant would have to support ten times the weight borne by a square centimetre of bone in a normal man. But this is just about the breaking strain of human bone, and Haldane concluded that the giants would break their thighs with every step they took, unless some radical alterations had been made in their internal construction, and different material used instead of bone.

Another mechanical difficulty is caused by the increased bending moments to which bone is subjected in a large animal. The amount of sag of a beam under its own weight varies as the square of the length of the beam, and in consequence the limbs, if they are to avoid breaking, must become thicker and shorter as the animal increases in size. Merely to increase the linear dimensions of a man is therefore not possible beyond a certain point, for his existence would become most precarious. His whole skeleton would have to be altered in the direction of shorter and more massive bones, with the inevitable requirement of more massive muscles to move them.

The giants of mythology were popularly supposed to be possessed of enormous strength. The strength of a muscle depends on two main factors, the number of fibres which can be brought to bear, and the pull exerted by each fibre. Now the number of fibres pulling depends on the cross-section of the muscle, and this, like that of the bones, increases with the square of the linear dimensions. But the weight which the muscles have to move increases with the cube of the same dimensions, so that on this account a very large man would be relatively more feeble than one of normal size. However, the pull of each fibre depends up to a point on the number of contractile units along its length, and the total number of these varies with the volume of the muscle, which increases with the cube of the linear dimensions. Large people are not, therefore, at so great a disadvantage as might be expected, and power more or less keeps pace with body size.

The surface area of the body, like the cross-section of the bones, only increases as the square of the linear dimensions. In practice this means that if all dimensions of one body are 10 per cent greater than those of another body of the same shape, the first body has 33 per cent more weight, but only 21 per cent more surface area. Now the rate at which heat is lost to the environment depends on the surface area exposed, and food intake and oxygen consumption must balance the loss of heat, otherwise death would ensue. A large person therefore needs less food per unit of weight than a small person does to keep him warm, and small people, like infants, are more vulnerable in situations where excessive heat loss is liable to occur, such as exposure to extreme cold.

In conformity with this principle it was once postulated that the largest animals should be found in cold climates, and the smallest in the tropics, but, although the proponents of this idea could point to polar bears and whales, the elephants and other large tropical animals such as the hippopotamus and the rhinoceros were conveniently forgotten. It was later urged that within a given genus the species enlarged in size as one progressed away from the tropics. There was more to commend this argument, but there are exceptions—for example, the otters tend to become smaller as they grow colder. The relation of human size to climate will be discussed in CHAPTER 7, but it may be said here that any systematic effect of this kind is outweighed by genetic influences and other factors such as diet. There is, however, some evidence that body *shape* may be determined in part by climatic surroundings [p. 129].

It is interesting to examine the devices which the larger animals have been obliged to adopt in order to increase the area of vitally important surfaces to correspond with the increased weight, and hence increased metabolic activity, of the body. The absorptive mucous membrane of the small intestine, for example, is thrown into circular folds, upon each of which have been developed large numbers of finger-like villi (the same device is used in the household bath-towel). The lungs have developed an enormous surface area by spreading their capillaries out on the walls of the alveoli; the cortex of the brain has been thrown into folds. These modifications parallel on the gross scale the changes in shape which individual cells may adopt in order to avoid the restrictions imposed by the surface/volume ratio [p. 5], and, like the microscopic modifications, cannot be effective beyond a certain point.

These, then, are some of the factors which impose limits on the size of human bodies which have undergone no radical structural alterations. The next obvious step is to attempt to determine these limits. It is no great problem to obtain the average height and weight for a given population, but actual measurements of the extremes of human size are difficult to obtain. Excessive or minimal height is often part of a stock in trade; many dwarfs and giants capitalize their appearance in the entertainment world, and wild exaggerations have always been common. The mythical giants of the past tend to be of impossible size, such as Finn MacCool, who threw the Isle of Man across from Ireland, leaving behind Lough Neagh, the hole from which it was torn. Patrick Byrne [PLATE 3], a real Irish 'giant', was thought during his life to be very much taller than the 231 cm (7 ft 7 in) which his skeleton was found to measure when John Hunter, after many difficulties, succeeded in obtaining it for investigation. People over this height are undeniably rare, but Arey cites a height of 289 cm (9 ft 6 in) as the known extreme. The tallest man of whom there are authentic recent measurements was probably Robert Wadlow, an American who at his death at the age of 22 measured 271 cm (8 ft 10¾ in). Most such cases of giantism are due to disorders of the pituitary gland [p. 188].

The causes of dwarfism are much more variable, but again many dwarfs suffer from pituitary deficiency. Perhaps the most famous dwarf was General Tom Thumb, who was exploited by Barnum, but many smaller dwarfs have been recorded, including a midget 48 cm (1 ft 7 in) tall at the age of 18 years and weighing 5·45 kg (12 lb). There were two siblings, one of whom measured 56 cm (1 ft 10 in) and the other 84 cm (2 ft 9 in); the father was 188 cm (6 ft 2 in) in height and the mother 168 cm (5 ft 6 in).

Even if we are satisfied to have obtained an estimate of the probable extremes of the range of human size, we are still faced with the problem (so common in all biological measurements) of defining the limits of what may be considered 'normal' or 'abnormal'. Dinka women, for example, would be considered extremely if not abnormally tall in Britain, whereas pygmies would be thought of as dwarfs. But when is a member of a specified race to be considered a giant or a dwarf? However the categorization is made, there is always an overlap zone in which an individual may be thought of as either a very small 'normal' or a dwarf, a very large 'normal' or a giant.

Within the 'normal' range there are many large and many small people, and there are advantages and disadvantages in being either. Many of the drawbacks to being excessively large derive from the fact that large people have not only a greater inertia, but also a greater momentum than smaller ones; it follows that as a rule small people have greater agility. As D'Arcy Thompson said: 'Among animals we shall see, without the help of mathematics or of physics, how small birds and beasts are quick and agile, how slower and sedater movements come with larger size, and how exaggerated bulk brings with it a certain clumsiness, a certain inefficiency, an element of risk and hazard, a preponderance of disadvantage.'

Nevertheless the maximal speed of which a large person is capable is very similar to the maximal speed of a small person. A. V. Hill, who investigated this problem, pointed out that 'if one animal is 1000 times as heavy as another, its linear dimensions will be 10 times as great, it will take 10 times as long for one movement, but since that movement is 10 times as great, its linear speed over the ground will be the same'. Large animals actually have an advantage in staying power: Hill calculated that it would take 10 times as long for the larger animal to become exhausted as it would for the smaller animal when both were exerting a maximal effort. To quote D'Arcy Thompson again: 'in the Oxford and Cambridge boat race it is prudent and judicious to bet on the heavier crew'.

There are other difficulties about being large in our present civilization. Although large people can see over hedges and are better off at football matches and other ceremonial occasions, they have enormous trouble with the apparatus of everyday life. Kitchens are still being designed for the undernourished dwarfs of the Victorian age, and even quite small giants who are rash enough to help with the washing-up may develop slipped intervertebral discs and other postural disabilities. Rakes, brooms, and other garden tools are designed for small people, and lawnmowers have handles so placed as to make it impossible for a tall man to push them unless he adopts a painful crouched position. Tall women have great difficulty in obtaining ready-to-wear clothes and shoes, and most coat-hangers are made for dwarfs. Hotel beds in many places are still provided on the assumption that people over 170 cm (5 ft 7 in) in height do not exist, and the seating in cars, cinemas, schools, and other public places is as a rule atrociously designed, even to the extent that school furniture has been suspected of contributing to the production of spinal deformities [p. 136].

If, as seems possible, Western man is at present slowly increasing in size and weight, it follows that more attention will have to be paid in future to the measurements of standard domestic constructions. It is also possible that the composition of the human body may eventually alter in response to the increase in size. It is a general principle of biology that in all except the simplest organisms growth of the various components of the body is differential [p. 68], so that, as growth occurs, all parts of the body do not increase proportionately. If an increase in the size of a given animal occurs in the course of evolution, it is common for differential growth to produce structural and functional changes in its organs and tissues, and, indeed, this seems to be one of the basic methods of natural selection. The changes which might occur in the human body in response to an increase in size are impossible to predict, and afford an interesting field for writers of science fiction.

3

Growth of Tissues

'It is not growing like a tree
In bulk, doth make men better be'

JONSON

CONNECTIVE TISSUE

Very little is known about the growth of connective tissue, but something is beginning to be learned about the deposition of collagen fibres from studies on tendons. It appears that the thickness of a tendon (and thus the number of collagen fibres developed in it) may depend on the average severity and duration of the stresses to which it is subjected. There is no strict relationship between the size of a muscle and the size of its tendon, and if a muscle belly is removed from a baby animal, leaving the tendon behind, the tendon, which is subjected only to incidental passive tensions during movement of adjacent muscles, may reach as much as 85 per cent of its normal adult diameter. But if the incidental movements are eliminated as far as possible by fixing the adjacent joints, the isolated tendon may not grow beyond its size at operation.

It has for a long time been supposed that collagen fibres are laid down in connective tissue along the lines of stress to which the tissue is subjected, and it is probable that the weight of collagen in any given situation depends on the magnitude of the local stresses. In a healing wound [p. 171] it has been experimentally shown that the directional posture adopted by the fibroblasts is determined by the tensional forces acting on the wound. Such a mechanism also accounts for the development of the retention bands which bind down tendons, the pulleys which alter their direction of action, and the ligaments which restrain excessive movements.

There was once some argument as to whether fat 'grows' in the same sense as the other tissues. The fat of the body is contained in connective tissue cells which are initially indistinguishable from fibroblasts, and it might thus be maintained that fat is merely a chemical deposit taken up by connective tissue cells in the course of their life, and that there are no such things as 'fat cells', and no such thing as 'adipose tissue'.

But fat does not occur in all parts of the body: for example, the subcutaneous tissue of the eyelids, the external ear, the nose, the back of the hand, and the scrotum contain very little fat indeed. On the other hand, fat is very commonly stored in the subcutaneous tissue elsewhere, and in the omenta of the peritoneum. This has been taken to mean that fat cells constitute a specific kind of connective tissue with a definite distribution, and this view is supported by the fact that if connective tissue taken from a site where fat will normally accumulate is transplanted to another site which is normally fat-free, it will still form adipose tissue in its new surroundings.

It is now possible to determine both the number and the size of the fat cells ('adipocytes') in a sample of fat, and there is evidence that in rats the number of such cells increases in the early stages of growth, and that the number can be permanently increased by overfeeding at this stage. However, after the animal has attained skeletal maturity (about 15 weeks of age) there is no further increase in number, and fat depots grow purely by cellular enlargement. If the rat is now starved, or if overfeeding is continued, the number of adipocytes remains unchanged but the size of the individual cells alters.

But even if fat is a deposit in a specific kind of tissue, it is still in rather a different category from muscle, bone, etc., when it comes to growth, for the fat deposits in the body are largely a means of storing energy, and at any stage during growth they can come and go according to the nutrition of the individual. Not all such stores of fat are equally labile. Thus, fat in the subcutaneous tissue and in the omenta is readily drawn upon in case of need, but other deposits, such as those in close relation to viscera like the heart and kidneys, are kept in being even in quite severe malnutrition.

Fat appears in the subcutaneous tissues of the fetus about the sixth month of fetal life, and is plentiful in this situation in the new-born baby, probably as an insulation against cooling [p. 121]. It accounts for about 25 per cent of the total weight of the infant. Characteristic pads of fat at the inner sides of the soles of the feet may give the impression that the baby is flat-footed; they disappear later. There is also a 'suctorial pad' of fat in the cheek, which is supposed to aid in sucking, though the mechanism is speculative. There is an accumulation of brown fat in the neck, round the kidneys and in the region of the scapulae; this apparently helps in temperature regulation, by acting as a store of rapidly usable energy. During the first 10 years of life this brown fat continues to be widely distributed, but thereafter it gradually disappears from most areas, though it persists, even into old age, round the kidneys, the suprarenals, and the aorta. The infant has little fat in several places where it is prominent in the adult; the omenta are small and undeveloped, and there is little fat round the kidneys.

The fat content of the body increases considerably between birth and the age of six months, but the rate of increase rapidly falls off in the last six months of the first year. After this the fat content actually *decreases* till the age of 6 or 8 years, by which time the thickness of the subcutaneous tissue is approximately half what it was at the age of 1 year. The decrease is less in girls than in boys, so that after one year of age girls are fatter than boys of the same age.

Subsequently the body fat begins to increase again, and many children put on excess fat just before the adolescent spurt; this may lead to emotional problems. During the male spurt the fat on the limbs decreases, and is not gained back until the late twenties. In girls there is no such decrease, though there may be a temporary interruption of the increase. In both sexes fat on the trunk continues to increase fairly steadily, but in girls additional fat is laid down in the secondary sexual distribution [FIG. 19].

Just as overfeeding rats in early life can have a lasting effect on body weight and fat cellularity, so in human babies it has been suspected that excessive food intake in the first 6 months of life could be associated with an obesity later. The rate of multiplication of adipocytes can be most affected by overnutrition before 1 year of age, and it seemed possible that in the first year a fixed complement of adipocytes is achieved, which cannot be subsequently reduced by dieting. All obese individuals show an in-

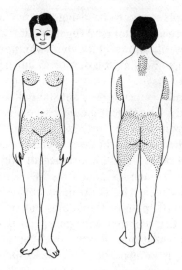

FIG. 19. Sites of deposition of fat in the female body at puberty.

crease in the average *size* of adipocytes, but only some of them have an increased *number* of these cells; it was suggested that the latter group were those who had an early onset of obesity.

This could be a serious problem. A survey of 300 infants in Worcestershire in 1972 found that 16·7 per cent were obese and a further 27·7 per cent were over the 90th centile standard for weight. This finding has been associated with the current habit of adding solids to the diet very early, while continuing to give the same amount of milk.

But there is also a high correlation between obesity in the parents and obesity in the children. This indicates a genetic factor of considerable importance, since there is no such correlation between the weight of adopted children and their adoptive parents. If obesity is taken as a deviation of 20 per cent or more from 'ideal' weight (p. 37), there is estimated to be a 70 per cent chance of becoming obese if both parents are obese, a 50 per cent chance if one parent is obese, and a 10 per cent chance if both are of normal weight.

There is some evidence that in the United States at least the biochemical composition of the body fat may be altering; the proportion of stearic and oleic acids seems to be falling and the proportion of linoleic acid to be rising. These changes are probably due to an increased intake of vegetable fats and a decrease in the consumption of animal fats, but whether this is due to commercial factors affecting the diet or to nutritional propaganda is not yet known.

Connective tissue makes a very substantial contribution to the weight of the body. The subcutaneous tissue alone accounts for as much as a quarter of the birth weight, and in the adult it contributes 6–7 per cent of the total weight in males and almost double this in females. The importance of connective tissue in relation to height is much less, and only two regions are involved, the scalp and the heel. The scalp need not detain us, though it does appear to participate in the adolescent spurt, but the fibro-fatty pad of the heel, together with the specially thickened skin in this region, may constitute a layer of 2 cm or more in depth. Virtually nothing is known about the development or growth of such pads.

At birth the bursae under tendons and muscles are well developed, but those under the skin may or may not be present.

SKIN

The skin at birth is relatively thin and easily damaged; it weighs about 4 per cent of the total body weight in the new-born, and about 6 per cent in the adult. Since there are about 800 square centimetres of skin per kilogram of body weight in the infant and only about 300 square centimetres in the adult, the disparity in thickness is greater than these percentages would suggest at first sight. Sebaceous glands are present in the fetus before birth, and their secretion is the main constituent of the vernix caseosa, a cheesy material covering the surface of the new-born baby and filling the external auditory meati. The sebaceous glands are relatively inactive for some time after birth, and they grow very rapidly during the adolescent spurt, at which time blackheads, pimples, and acne are rife. These are associated with the profound hormonal changes which occur at puberty; eunuchs do not suffer from acne.

The sweat glands in the new-born baby are immature and function poorly; this is important because of the effect on the regulation of the body temperature.

In the later stages of fetal life there is a generalized covering of delicate hairs known as the lanugo, but these are mostly lost before birth, and after a few weeks are replaced by secondary hairs, also fine, known as the vellus. The persistent lanugo hairs on the scalp and eyebrows are shed after a few months and replaced by thicker 'terminal' hairs. At adolescence the coarse secondary sexual hair appears, under the influence of the secretions of the gonads [p. 105].

No new hair follicles are formed after birth, except in response to local trauma, and the successive changes in the type of hair occur in the same pre-existing follicle. The transition from vellus hairs to terminal hairs may be abrupt or gradual; in the latter case successive generations of hairs in the same follicle form a transition. Each follicle is specific for its own type of terminal hair, and will continue to produce this hair if transplanted elsewhere; thus, hair from the scalp will not change its character if transplanted to the pubic region and vice versa. Occasionally the hair pattern can be disturbed by certain follicles producing a type of hair inappropriate to the age and sex of the patient.

The skin and its appendages are a convenient site in which to observe growth in the adult, and further details will be given in CHAPTER 8.

MUSCLE AND TENDON

The composition of muscle varies with age; in the fetus it contains a great deal of water and much intercellular matrix, but after birth both are considerably reduced as the cells grow in size by accumulating cytoplasm [p. 4]. There is curiously little information on the numbers of fibres in a given muscle at different ages. In 1898, counts made on the human sartorius muscle indicated that there is no multiplication of fibres in this muscle after the 4th or 5th month of fetal life, and that the subsequent increase in size is due to enlargement of the existing fibres. It seems to be generally agreed that this finding applies also to the other muscles of the human body. Nevertheless, it has been observed that in rats and mice the number of fibres in a given muscle may increase with age. In some marsupials this increase is accompanied by a decrease in the diameter of the fibres, and this has been interpreted as meaning that muscle fibres may split longitudinally as they grow, so forming two daughter fibres. In contrast to the fibres, muscle fibrils remain constant in diameter as the muscle grows, and it follows that their number increases. They grow in length at their ends. The growth of the connective tissue components of a muscle is maximal in the vicinity of the junction of the tendon with the muscle. Mitotic figures have been demonstrated in skeletal muscle after birth, but electron microscopy shows that these cell divisions do not take place within the muscle fibres, but rather in the undifferentiated cells which are found outside them. These 'satellite' cells appear early in development, and their nuclei account for about 5–10 per cent of the total number in the muscle. They multiply to reproduce their kind and also to form muscle cells, which fuse together to increase the length of the muscle fibre. It is possible that there may be two distinct populations of satellite cells, one which perpetuates the race of satellite cells and the other which produces new muscle cells.

The total mass of muscle in the body can be estimated from the amount of creatinine excreted in the urine, since the conversion of creatine to creatinine takes place only in muscle. The number of nuclei in a known volume of muscle removed at biopsy can be estimated by finding the amount of desoxyribonucleic acid (DNA) it contains, since the amount of DNA per nucleus is constant. From these two sets of figures the number of muscle nuclei in the body can be obtained, and it is found that in boys the number increases 14-fold between birth and maturity; in girls the increase is 10-fold. The size of the muscle fibres reaches a maximum in girls about the age of 10, and not until about the age of 14 do the boys catch up. It seems clear that the increase in length of muscle fibres which was always known to accompany growth has now been shown to be paralleled by an increase in the number of nuclei in the fibres. The question regarding the possible increase in the number of muscle fibres remains unsolved.

A potent stimulus to the growth of a muscle is the separation of its attachments as the skeleton grows, and it has been suggested that the length of a muscle fibre is the direct consequence of the range of movement it is called upon to perform. Common sense would certainly indicate that this is so, for if the fibres were too short they would rupture under the strain of a full movement, and if they were too long they could not exercise their full potential for contraction. The stimulus for the formation of a tendon is probably the pull of the muscle rudiment on undifferentiated connective tissue [p. 46], and certainly the repair of a tendon from which a segment has been removed does not occur if the muscle is prevented from exercising tension on the cut tendon.

An interesting experiment is to release the tibialis anterior of an experimental animal from the extensor retinaculum, thus making the distance between the bony attachments of the muscle shorter. The result of this change is that the growth of the tendon is retarded. But because the muscle fibres have now to contract over a greater range in order to produce the same amount of movement at the ankle joint, there is an actual *increase* in the length of the muscle belly.

Replacement of muscle by tendon ('tendinification') sometimes appears to be the result of limitation of movement. In animals such as the baboon the coccygeus passes from the ischial spine to the mobile coccygeal vertebrae, and is a fleshy muscle throughout; in man, where the upper part of the muscle is attached to the immovable sacrum, this portion is converted into fibrous tissue.

If the normal opponents of a muscle are paralysed, the muscle fails to grow properly, and it may be that the full development of a muscle is dependent on the progressive rise in the tension exerted on it by its antagonists.

Many of the fibres in the muscles of baby laboratory animals are relatively rich in the enzyme myosin ATPase, which enables them to contract rapidly. As the animal grows, the proportion of such fibres falls, particularly in the postural muscles, which have to maintain the increasing body weight against gravity for progressively longer periods. It has not yet been established whether the relative proportions of these kinds of fibres in human muscles change between childhood and adult life.

The growth of muscle at adolescence is a big factor in the adolescent spurt in weight, which affects all components of the body, except, perhaps, the central nervous system. Before the time of the adolescent spurt, the strength of boys and girls shows little difference, but after the spurt males have the advantage in grip, arm pull, etc. [FIG. 20]. The curve for the increase in strength lags about a year behind the curve for the increase in weight of the body; hence there is a limited sense in which children may be said to 'outgrow their strength'. The maximum

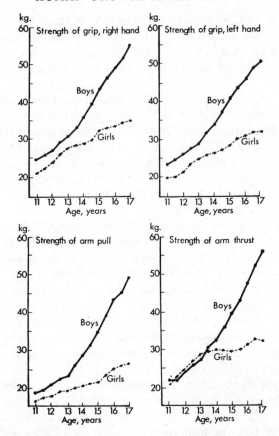

Fig. 20. Strength of grip, arm pull, and arm thrust. Mixed longitudinal data. Note the divergence of the curves for boys away from the curves for girls at the time of the adolescent spurt. (From Tanner, J. M. (1962) *Growth at Adolescence*, 2nd edn, Oxford, Blackwell Scientific Publications Ltd., by kind permission of the author and publishers.)

muscular strength is usually attained between 25 and 30 years of age, and after this there is a gradual and small falling off in both the speed and power of contraction. After 50, the muscles begin to lose fibres and hence become considerably reduced in size [p. 215].

It has already been said that in the infant the muscles account for something like a quarter of the total weight, but that in the adult this figure rises to about 40 per cent [p. 39]. The muscles of the head, trunk and upper limbs are relatively heavier in the infant because of the poor development of the lower limbs; in the adult about 55 per cent of the weight of the musculature is accounted for by the lower limb muscles. The respiratory muscles and those of facial expression are well developed at birth to provide for the vital functions of breathing and sucking, but others, such as the levator ani, have been described by Crelin as having 'the consistency of moist tissue paper'. The weight of the musculature can be increased, in both children and adults, by exercise [p. 157].

A good deal less is known about visceral and cardiac muscle than is known about somatic muscle. It appears that the cells of visceral muscle increase both in number and in size, but that the latter process is the more important; it has been calculated that visceral muscle can increase in bulk by a factor of eight times merely by the enlargement of its fibres. Cardiac muscle also seems to increase in bulk mainly by enlargement of existing fibres.

NERVOUS TISSUE

It is believed that after the sixth month of fetal life no new nerve cells are formed [p. 10], and that subsequent growth of the nervous system depends on an increase in the size of the existing cell bodies, enlargement, ramification and myelination of their processes, and auxetic and multiplicative growth of the specialized connective tissue cells which support the nervous elements.

Once formed, a nerve cell can increase in mass up to 200 000 times, most of the addition being to the processes of the cell, which may come to contain as much as a thousand times the amount of material contained in the cell body. The diameter of the myelinated nerve fibres in peripheral nerve trunks increases during growth very considerably, and the nerve cell is rich in ribonucleic acid (RNA), which is used for the heavy task of forming cytoplasm to be pushed outwards into these fibres. In tissue culture the growth of nerve cells both in size and in number is specifically stimulated by a substance known as nerve growth factor, which can be isolated from a number of curious places, such as the salivary glands of mice.

Myelination [FIG. 21] is a progressive process which is far from complete at birth [p. 81]. In the ventral roots of the spinal nerves of the rat the first fibres to myelinate have very different diameters, and there seems to be no critical size or range of sizes at which myelination begins. As the length of the myelinated fibres increases, the total number of nodes of Ranvier appears to remain constant. It follows that the length of the internodal

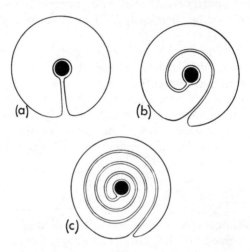

FIG. 21. Scheme of process of myelination.

(a) Cross-section of a nerve fibre embedded in the cytoplasm of a Schwann cell. The gutter along which the fibre was engulfed by the Schwann cell still remains open to the surface.

(b) The Schwann cell and the fibre rotate relative to each other, so winding the adjacent surfaces of the Schwann cell membrane round the fibre.

(c) A further stage of the winding process, showing several double layers of membrane surrounding the fibre. These tightly wound wrappings fuse together to form the myelin sheath of the fibre.

segments increases with growth. After a fibre has been surrounded with myelin, it continues to require further myelination on account of growth both in length and in diameter. In the central nervous system progressively more complex connexions and assemblies are formed between nerve cells as growth progresses. The spiral winding mechanism of myelination shown in FIGURE 21 cannot apply in the central nervous system, where myelination is effected by the connective tissue oligodendrocytes, since a single oligodendrocyte can form myelin sheaths round a number of separate fibres simultaneously.

CARTILAGE AND BONE

The skeleton contributes about 14 per cent to the weight of the adult body, and, along with the fibrocartilaginous intervertebral discs, something of the order of 97–98 per cent of the total height. (The remaining 2–3 per cent is due to the thickness of the scalp and the fibro-fatty pad of the heel, together with the thickness of the articular cartilages in the lower limb and in the atlanto-occipital joint.)

A great deal more is known about the growth of bone than about the growth of any other tissue, because of the accessibility of bone to investigation in the living person by means of radiology. It is therefore possible to discuss the processes involved in some detail.

Formation and Growth of Cartilage

Towards the end of the first month of fetal life the embryonic connective tissue in the region of the future skeleton begins to show signs of differentiation. The primitive connective tissue cells become more closely packed, and lose the processes which up to now have radiated from them. Later they begin to lay down a matrix rich in chondroitin sulphate (cartilage is about 70 per cent water, and, of the residue, 80 per cent is collagen and sulphated polysaccharides), and this gradually separates the cells from each other again. The cells round the developing cartilage begin to form two layers: in the outer one they differentiate into fibroblasts, and coincident with this collagen is laid down [p. 11]. The cells of the inner layer remain more or less undifferentiated and capable of division to produce cartilage cells. The two layers together are called the perichondrium.

The newly-formed cartilage grows larger in two ways, by interstitial and by appositional growth [p. 3]. Interstitial growth occurs because the cells in the centre of the developing mass do not immediately lose their power to divide. The new cells which they form join with the older ones in laying down more and more matrix, and the whole tissue grows rather as dough does in bread-making. Obviously this process depends on the matrix being pliable, and after a short while the matrix becomes too rigid for much interstitial growth to occur, and the second mechanism, appositional growth, takes over.

In this process the cells of the deeper layer of the perichondrium divide: some of the offspring remain as stem cells, and others differentiate, so that more and more cartilage is laid down on the surface of the existing mass. The differentiated cells then surround themselves with matrix in the usual way, and are in turn overlaid by successive new layers so that they gradually sink into the depths. As the mass increases in size, the surface area of the perichondrium expands to cover it, and an increasing number of stem cells is therefore required.

Ossification

During the second month of fetal life, bone formation begins. In a few bones, such as the clavicle and the bones of the vault of the skull, ossification begins directly in the connective tissue of the embryo, and such bones are said to be ossified 'in membrane'. The remaining bones of the skeleton are ossified 'in cartilage', which means that the same sort of process takes place after the connective tissue has become converted into a sort of cartilaginous template: the only difference is that cartilage bones have to go through this extra stage.

The points at which bone formation begins are known as 'primary centres' of ossification, and these centres appear in different bones at different times, the first usually being either the mandible or the clavicle, at about the fifth week of fetal life. In a membrane bone, the first sign of ossification is the penetration of a blood vessel to the spot, bringing with it specialized cells called osteoblasts and osteoclasts. Around the osteoblasts collagen fibres are deposited, and on these fibres calcium and other inorganic salts accumulate. The osteoclasts, which are large multinucleated cells, have the task of shaping the growing bone tissue by removing unwanted material. There has been a great deal of controversy about the exact methods employed by the osteoblasts and osteoclasts, and the matter is too complicated to go into in a book of this size. It is generally accepted, however, that osteoblasts are in some way concerned with bone deposition and osteoclasts with its removal.

In the primary centre of a cartilage bone, the first sign of ossification is that the cartilage cells swell up, and arrange themselves in columns. At the same time calcium salts are deposited in the matrix of the cartilage, so converting it into what is known as 'calcified cartilage'. A blood vessel now grows into the region from the perichondrium, bringing with it osteoblasts and osteoclasts as in the case of membrane bones. The osteoclasts remove some of the calcified cartilage, collagen fibres appear round the osteoblasts, and gradually the calcified cartilage is replaced by true bone.

Growth of Short Bones

The subsequent growth of a short bone such as one of the cuneiform bones of the foot can be illustrated by a diagram [FIG. 22]. In the centre of the cartilage, which is by now taking up the shape of the adult bone and is therefore known as the 'cartilaginous model', there is a steadily enlarging core of bone. At the same time the original cartilage is growing appositionally. A race develops between the two processes of cartilage deposition and

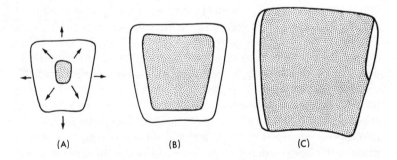

(A)   (B)   (C)

FIG. 22. Growth of short bone

(a) The core of newly formed bone in the centre of the cartilaginous 'model' increases in size at the expense of the cartilage. At the same time the cartilaginous model itself grows by activity of the perichondrium.

(b) Growth of new bone proceeds faster than growth of cartilage, so that a greater proportion of the developing structure is composed of bone.

(c) Bone formation has outstripped cartilage formation and the whole structure is now composed of bone with the exception of a large cartilaginous articular facet on one surface and a smaller one on the opposite surface. This is the adult condition for this particular bone, and growth now ceases.

bone formation, and continues until the adult stage of development is reached, at which time the bone formation 'catches up' with the deposition of cartilage so that the cartilage disappears, except for a thin rind which is left on the surface wherever the bone takes part in a joint with its neighbours. Because the perichondrium has come to surround bone and not cartilage, we now call it periosteum, but it retains the same two layers.

The perichondrium covering the cartilaginous joint surfaces disappears in the course of development. Such surfaces do not stop growing with the attainment of maturity, for they are exposed to considerable wear and tear which has to be made good [p. 161], but reparative growth in this region is necessarily interstitial, since there can be no contribution from the perichondrium. Similar considerations apply to the articular cartilages at the ends of the long bones, which may retain a thickness of 5–6 mm in the larger joints.

Growth of Long Bones

The growth of the long bones of the limbs is more complicated. The primary centre appears in the cartilaginous shaft of the future bone in exactly the same way as in a short bone. But, about the same time, bone begins to be formed on the surface of the shaft, due to the appearance of osteoblasts in the perichondrium, which now becomes the periosteum. The active cells in the deeper layer of the periosteum cause the shaft to grow in thickness. At the same time the bone formed in the primary centre is eroded away by osteoclasts, leaving a cavity which is not built in by osteoblasts, and forms a convenient storage place for the blood-forming cells and fat which constitute the bone marrow. As ossification proceeds, so the marrow cavity extends along the shaft of the bone and at the same time grows in diameter because of osteoclastic activity in its walls. The bone continues to grow thicker by appositional growth on the outside of the shaft, but as it does so the destruction on the inner surface keeps pace, and in this way the ratio of the total diameter of the shaft to the thickness of its walls is maintained more or less constant until maturity.

While these events are taking place, the expanded ends of the cartilage model grow, at first by interstitial, and later by appositional growth. As time goes on, other centres of ossification appear in the cartilage masses at the ends of the future bone [FIG. 23]. Such 'secondary centres' are, except for a few examples in reptiles and birds, confined to mammals; they are therefore not an essential part of bone growth in lower animals. In some sites two or three secondary centres may appear and eventually fuse together to form one large centre. A few secondary centres appear before birth, but the majority appear later, at a time

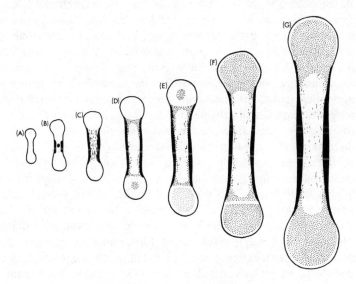

FIG. 23. Growth of long bone

(a) Cartilage model.

(b) Primary centre appears in the shaft and a collar of compact periosteal bone surrounds it.

(c) The shaft is ossified but the ends remain cartilaginous.

(d) The marrow cavity appears in the shaft and progressively enlarges. A secondary centre eventually appears at one end of the bone. The walls of the shaft are made of compact bone.

(e) Another secondary centre has appeared. The first has now formed an epiphysis of cancellous bone covered by cartilage and separated from the bony end of the shaft by a cartilaginous epiphyseal plate.

(f) The second epiphyseal plate is overwhelmed by the ossification proceeding into it from the shaft and has been converted into bone. Growth at this plate has ceased and the epiphysis is said to be 'closed'.

(g) The cartilaginous epiphyseal plate at the 'growing end' of the bone continues to grow in thickness for some time, but eventually it too is replaced by bone and growth in length comes to a halt. Throughout the whole process of growth in length, growth in thickness of the shaft has proceeded steadily by apposition from the periosteum.

characteristic of the individual bone; many do not begin their activity until puberty. There is no relationship between the time of appearance of the various centres and the volume of bone they have to produce before growth stops; it follows that some of them must progress much faster than others. The mass of bone formed by a secondary centre or an agglomeration of secondary centres is called an epiphysis, and, because of their calcium content, these epiphyses are readily seen in radiographs [PLATE 4], where they show as a dense shadow separated from the shaft by a translucent zone of still growing cartilage. This zone of cartilage is known as the 'epiphyseal plate', and it is a vital factor in the growth in length of the bone.

Cartilage cells in the epiphyseal plate increase its thickness by interstitial growth at the same time as ossification proceeding outwards from the shaft is destroying it and manufacturing bone. The result is that a race similar to that which occurs in the short bones takes place, and eventually the cartilaginous plate is overwhelmed, and becomes replaced by bone, which therefore unites the epiphyseal mass with the bony shaft. This 'closure' of the epiphysis terminates the growth in length of the bone, and in most of the long bones this event occurs at about the age of 16 to 18 in the female and 18 to 21 in the male. It is interesting that in rats some of the epiphyses of the long bones never close, but merely become inactive; such a quiescent zone of growth may be reactivated by various stimuli in fully adult animals. For example, rats can be made to grow at any time during their lives by injections of somatotrophin [p. 144]. This is not possible in man once the epiphyses have closed.

One of the earliest things to be discovered about the growth of a limb bone was that interstitial growth in length does not occur. This was shown by Stephen Hales, and subsequently by John Hunter, who made marks on the shaft of a bone, allowed the animal to grow, and subsequently found that the distance between the marks remained constant. Similar evidence can now be obtained by radiology after embedding two pellets of metal in the substance of a growing bone. Radiology is indeed the chief method of studying the growth of human bone, but other methods can be used with animals. For example, pigs readily eat the plant madder (*Rubia tinctoria*), which contains a compound similar to the dye alizarin. This is deposited where bone is being laid down and stains it pink. If an animal is given madder to eat for a known period, it can then be killed and the bones can be examined; all bone formed during the time of madder feeding will be stained pink, and will stand out from the normal white bone [FIG. 24]. In more recent years further observations have been made using tetracycline, which is deposited mainly in the periosteal growing region, in the epiphyseal plate, and in the region at the end of the shaft. Other workers have used radioisotopes such as ^{85}strontium.

In man, more growth always takes place at one end of a long bone than at the other, and the more active end is sometimes known as the 'growing end' of the bone. This is perhaps misleading, since both ends grow, but the term does serve to indicate the end at which an injury might result in more disturbance in the growth of the limb. There are important differences in the rates of growth of individual bones. Thus, the growing end of the

FIG. 24. Shaft of the tibia of a pig fed on madder. The madder was omitted from the food for thirty days before the animal was killed, and the bone formed during that time is white, while the previously formed bone is stained red (indicated by stippling). (Drawn from a photograph by Brash, J. C. (1934) *Edinb. med. J.*, *41*, 305–19; 363–87.)

femur puts on length roughly twice as fast as the growing end of the tibia. Injury to the epiphyseal plate at the lower end of the femur is thus much more serious in its effects on growth than damage to the plate at the upper end of the tibia. If the growth at one end of a long bone is experimentally prevented, there is an increase in the rate of growth at the other end, but so far there is no explanation for this observation.

Just why one end of a bone should grow faster than the other is not known: the various influences which determine bone growth [CHAPTER 7] would be likely to affect both ends equally, and there appears to be no histological or biochemical difference between the epiphyseal plates of the two ends. It is similarly curious that in all long bones except the fibula the secondary centres which appear last are the first to fuse with the shaft. Occasionally, as in the rat, fusion does not take place at all, and the epiphyseal plate remains visible on a radiograph as a radio-translucent line separating shaft from epiphysis. In such cases growth usually stops at or about the normal time.

As in most living things, the pressure exerted by the growth process is enormous, and it has been estimated that in order to stop growth at an epiphysis in a human limb it is necessary to apply a force of some 400 kg. Pressures less than this do not entirely stop growth, but they may impede it, and if the pressure is applied at an angle to the line of growth, deformities may result through the growth being misdirected. It has been suggested that certain deformities may be the result of faulty sitting or sleeping postures in small children; for example, that the prone knee-chest sleeping position may give rise to bow legs through pressure being applied at an angle to the epiphyses at the knee.

Growth is often not uniform at different places within the same epiphyseal plate. This helps to maintain stability by producing ridging and grooving of the bony surfaces, and is also responsible for the 'cupping' of some epiphyses, such as the composite one at the upper end of the humerus, which fits over the conical end of the shaft of the bone.

At the ends of the bones, where growth is rapid, spongy bone tends to be produced. This has a honeycomb structure, in which thin plates of bone criss-cross each other, enclosing in their meshes little cellular spaces in which is packed the marrow. The directions taken by these plates, which are called trabeculae, depend on the stresses thrown on the particular bone, and it seems that the collagen fibres, on which the inorganic salts are deposited, are laid down in such a way as to resist the pressure and tension lines of force in the bone. The most celebrated example in the body is afforded by the head and neck of the femur, in which the stresses can be compared to the stresses in a crane head [FIG. 25]. It appears that the trabeculae may at first be laid down at random, but that those which are inclined obliquely to the shearing forces in the bone are moved out of the way by these forces. Trabeculae which happen to lie in the direction of a pressure or tension line will be comparatively undisturbed, since along these lines there is no shearing force. Nevertheless, a genetic factor is also involved, since bone removed from the body and grown in nutrient medium will also show similarly oriented trabeculae.

In the shaft of the bone, where growth is relatively slow, the bone produced by accretion is compact and dense. It has in fact been suggested that rapid bone growth always produces spongy bone, and slow growth produces compact bone. Whatever the reason, the compact bone of the shafts of the long bones is thickest where it has to withstand the maximum stresses; in general, this is about the middle of the shaft.

Moulding and rearrangement is not confined to the interior of

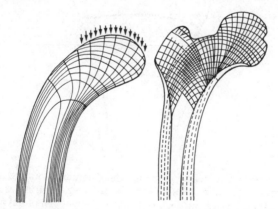

FIG. 25. Lines of stress in cancellous bone. The upper end of the femur (right) compared with lines of stress in the head of a crane (left). (From Thompson, D'Arcy W. (1942) *Growth and Form*, 2nd edn, London, Cambridge University Press, by kind permission of the publishers.)

the growing bone. As growth proceeds, it is necessary that some of the existing surface material should be removed if the characteristic shape of the bone is to be preserved [FIG. 26], and this is done by the osteoclasts. If anything goes wrong with the moulding process, the ends of the bones will be clumsy, thickened, and inadequately suited for their purpose. Some of the external moulding of a bone depends on the pressures and stresses exerted by the local soft tissues—for example, the pull of muscles attached to the bone, or the pressure of a neighbouring nerve trunk. But these effects can only be considered as a kind of 'fine adjustment', and a bone grown in tissue culture, with no such influences acting on it, will still assume a shape not too different from the normal adult configuration.

Unlike the bone itself, the fibrous periosteum which covers it increases in surface area by general interstitial growth, and therefore some sliding has to take place between the two structures. Since the muscles have their primary attachment to the periosteum, which is in turn attached to the bone by Sharpey's fibres, it follows that considerable local readjustments must occur in the case of muscles attached where much sliding takes place. The

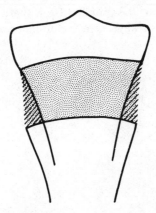

FIG. 26. External modelling of bone. The diagram shows the upper end of the tibia, the outline of the adult bone being superimposed on the outline of the 14-year-old stage of its development. The total amount of bone laid down by the epiphyseal plate during the time interval is shown by the stippling, and the amount of this bone which has to be removed in order to keep the contour of the bone constant is shown by the cross-hatching.)

attachments of muscles have also to be rearranged because of the continuous process of moulding of the prominences or depressions to which they are fixed.

Bone also undergoes internal reconstruction during growth. The distribution and arrangement of the trabeculae alter continuously to maintain the pattern induced by external stresses, and the Haversian systems of compact bone which is undergoing moulding are removed and reorganized.

Once bone has been formed it is by no means inactive; it is constantly being destroyed and reformed, and the picture presented by radio-isotope studies is one of an intensely busy tissue, in which the maximal rate of structural change occurs round about the age of $2\frac{1}{2}$ years. In young adults there is a relatively low turnover of material in bone, and later in life the amount of destruction gradually comes to predominate over the amount of replacement, so that bone tends to become rarified and weakened in old age [p. 224].

Growth of Movable Joints

The articular cartilage of synovial joints shows some response to stresses put upon it. For example, if a joint for any reason acquires an increased range of movement, the cartilage tends to extend in the required direction; this may occur even in adults. Conversely, if movement is restricted, part of the cartilage may become converted into fibrous tissue, and the articular surfaces may actually fuse together in places.

4

Growth of Systems

'I watched my foolish heart expand'

BROWNING

PATTERNS OF GROWTH

If the size of the body at different ages is plotted as a percentage of its size at the age of 20 years (arbitrarily representing maturity), a curve similar to the solid line in FIGURE 27 is obtained. The respiratory, digestive, and excretory systems follow very similar

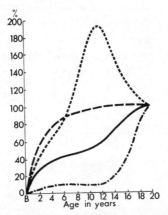

FIG. 27. The growth of systems and organs. Percentages of adult weight attained are plotted against age in years. (See text for description.) (Redrawn from Scammon, R. E., The measurement of the body in childhood, in Harris, Jackson, Paterson and Scammon (1930) *The Measurement of Man.* University of Minnesota Press, Minneapolis, University of Minnesota, by kind permission of the publishers.)

curves, and thus their relative sizes remain more or less constant during growth. It is also clear that the skeleton must behave similarly, since it is responsible for stature, and this is correlated with weight [p. 37]. Other systems, however, follow different growth curves.

Lymphoid Pattern

The dotted line in FIGURE 27 shows the growth of the lymphoid structures such as the tonsils, the thymus, and the lymph nodes. All these are prominent in the young child, grow rapidly in early childhood, and reach their maximum size around the age of puberty. After this the lymphoid tissue begins to degenerate and to be removed, and by the age of 20 it is much less in amount than it was at the age of 12 or 13 years.

Neural Pattern

The central nervous system and the organs of special sense, together with the skull which contains them, follow a curve in which there is very rapid growth in the early stages of childhood—so rapid that the structures involved reach about 90 per cent of their adult size by the age of 5 or 6 years. At 12 a child has a head which is almost as large as that of his father, and can usually wear his father's hat. From this time on, the growth of the nervous system is much slower than that of the rest of the body [FIG. 27, dashed line].

Gonadal Pattern

A third pattern of growth is followed by the gonads and the external genitalia, which develop very slowly in the early stages of growth, but at puberty begin to grow much faster than the rest of the body [FIG. 27, dashed and dotted line].

Uterine Pattern

Finally, the uterus and the suprarenal glands share a fourth pattern, being relatively large at birth, actually *losing* weight rapidly, and not regaining their birth weights until just before puberty [pp. 94, 97].

Not only do the different systems and organs follow different patterns of growth, they also increase in absolute size and weight

TABLE 3

WEIGHT INCREASES IN ORGANS AND TISSUES FROM BIRTH TO MATURITY

INCREASE (APPROXIMATE)	
30–40 times	Somatic muscles
	External genital organs
	Testicles
	Pancreas
25–30 times	Uterus
20–25 times	Body as a whole
	Skeleton
	Respiratory system
15–20 times	Heart
	Liver
	Lymphatic system
	Ovaries
10–15 times	Thyroid gland
	Kidneys
5–10 times	Pituitary gland
< 5 times	Nervous system
	Suprarenal glands

by very different amounts. TABLE 3 shows the approximate increase in weight from birth to maturity of a selection of organs and systems.

THE SKELETON

The Limbs

At birth primary centres of ossification are present in all the long bones, and one or two secondary centres have appeared; the secondary centre in the lower end of the femur is used as a legal index of the full maturity of the infant. Other secondary centres often present at birth include those for the heads of the femur and humerus, and that for the upper end of the tibia.

Primary centres in the limb girdles are established early in intrauterine life, and by the time of birth have produced a considerable amount of bone [FIG. 28], but no secondary centres have yet appeared. In the tarsus, centres for the calcaneus, the talus, and usually the cuboid are present at birth, but in the carpus the first bone to begin ossification (usually the capitate) does not as a rule do so until some time during the first year after

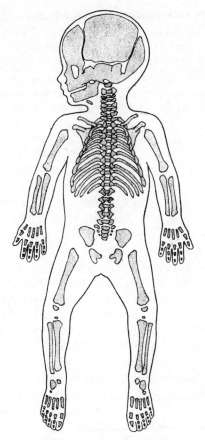

FIG. 28. Ossification of the skeleton in a new-born baby. Note the absence of any centres in the wrists, the separate centres for the components of the hip bones, and the lack of any secondary centres for the long bones except the lower end of the femur and the upper end of the tibia.

birth. Primary ossification in the metacarpus, metatarsus, and phalanges begins during fetal life and the shafts are well ossified at birth.

The times of appearance of secondary centres in the limb bones are scattered over a long period, among the last group to appear being some as late as the fourteenth or fifteenth year. The bone which they produce fuses with that produced by the primary centres at times varying from 16 years or so in the female elbow region to about 21 years for the male wrist region. The clavicle is usually the last long bone to stop growing, sometimes as late as the twenty-fifth year. The 'growing ends' of the long bones of the upper limb are the upper end of the humerus and the lower ends of the radius and ulna; in the lower limb the position is reversed, and the lower end of the femur and the upper end of the tibia undertake most of the growth in length.

The presence or absence of an epiphyseal line in a given situation at a given age is of considerable importance in the interpretation of radiographs taken after injury, for a perfectly normal line may be mistaken by the inexperienced for a fracture. Again, violence applied to a given region may result in displacement of an epiphysis if the patient is a child, while similar violence in an adult can only produce either a fracture or a dislocation of the adjacent joint.

At birth the bones of the pelvis and lower limbs are less advanced in relation to their ultimate size than those of the upper limb and shoulder girdle; they therefore have to 'catch up' by growing at a greater rate. Some of the most marked alterations in the skeleton occur in the pelvis. At birth the ilia and the sacrum are more upright than in the adult, and the subpubic angle is more acute, so that the cavity of the pelvis is not only very small, but also more funnel-shaped. After birth the growth of the pelvis keeps pace with the growth of the lower limbs. With the beginning of walking [p. 133], the curvature of the sacrum increases, the ilia become thicker and stronger, and the acetabula become deeper. Later, at puberty, the characteristic sex differences in the pelvis become manifest [p. 124].

The articular surfaces of the pubic bones show characteristic progressive modifications throughout adult life.

Skull and Mandible

If the growth of the limb bones is complicated, that of the skull is more complicated still. At birth the central nervous system is relatively very large [p. 81], and consequently the vault of the skull, the part which contains the brain, has to be large also; its capacity is about 400 ml (adult 1300–1500 ml). After birth it follows the very rapid further growth of the brain; by the age of 2 years its capacity is approximately 950 ml, and its circumference has increased from about 33 cm to about 47 cm. The cranial vault is ossified in membrane, and at birth the bones which compose it are separated by gaps filled with fibrous connective tissue; these allow the bones to slide over each other to some extent as they pass through the narrow birth passages of the mother. The largest of these gaps may be felt in the top of the skull between the frontal and parietal bones; it becomes tense when the baby cries. This anterior fontanelle, as it is called, is about 3 cm across, and is not filled in by ossification until the second year of life; other, shorter-lived, fontanelles lie between other components of the cranial vault [FIG. 29].

When the ossification process extends into the fibrous connective tissue, the individual bones of the vault come into relation with each other along a series of fibrous joints known as sutures. The edges of the bones are covered by a growing layer of osteoblasts and osteoclasts, and growth of the skull takes place at the sutures, at first very rapidly, in association with the growth of the enclosed brain, and later more slowly. As the bones grow laterally, so additional bone is deposited on the outer surface of the vault, by the usual process of apposition from the pericranium, as the periosteum is called in this situation. At the same time, osteoclasts remove bone from the inside of the vault, so that the brain cavity grows in much the same way as the marrow cavity of a long bone [FIG. 30]. As growth of the vault continues, the remodelling processes inherent in bone growth convert the original single layer of bone into two layers of compact bone with a layer of spongy bone in between. The complexity of accretional growth and remodelling in such a system can be imagined.

After the first few years of life, many of the sutures of the vault come to interlock with each other like pieces in a jigsaw puzzle, and form squiggly serrated lines; from this time on, growth at the sutures becomes a less important factor in increasing size, and it stops altogether about the time of puberty. Later still the sutures are obliterated, by extension of ossification into the sutural joints, and the individual bones become fused to each other. This obliteration may begin on the inner surface of the vault about the age of 25–30 years, and becomes visible on the outside of the skull some ten years later, the sagittal suture usually fusing first. As with the obliteration of the epiphyseal lines in the long bones, there is great variation in the time at which the sutures close, but

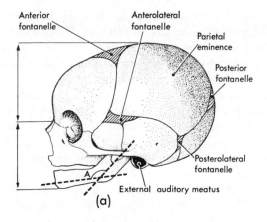

(a)

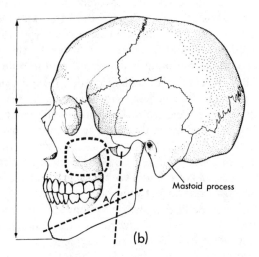

(b)

Fig. 29. Growth of skull

(a) Skull at birth. Notice (i) the fontanelles; (ii) the smooth non-serrated suture lines; (iii) the absence of a mastoid process; (iv) the obliquity of the angle of the mandible; (v) the relatively small size of the face.

(b) Skull of adult. Notice (i) absence of fontanelles; (ii) certain suture lines are now serrated; (iii) strong mastoid process, (iv) more upright mandible; (v) great increase in the vertical diameter of the face due to the eruption of teeth and the development of the jaws. The position of the maxillary sinus is indicated by the dotted line.

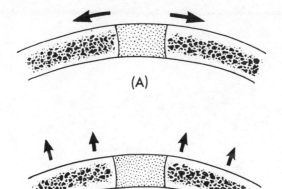

FIG. 30. Growth of vault of skull

(a) Growth at suture. The fibrous tissue separating the bones increases in amount and is at the same time invaded by ossification spreading from the adjacent bones. In this way the bones are pushed further apart and lateral growth occurs.

(b) Appositional growth. New bone is added to the outer surface of the skull while excavation takes place inside, so maintaining the thickness of the vault relative to the total size of the skull and allowing the brain to expand concurrently.

in general it may be said that partial obliteration of the sagittal suture indicates an age in the early 30s, of the coronal suture an age of about 40, and of the lambdoidal suture an age of about 50.

As growth proceeds, not only does the size of the cranial vault increase, but its contour alters. For example, the parietal eminence [FIG. 29] which is well marked in the infant skull, becomes much less prominent in the skull of an adult, because surface apposition is more vigorous on the surrounding bone than on the eminence itself.

While the growth of the skull vault and orbit is related to the growth of the nervous system, that of the remainder of the face and the base of the skull is more closely associated with the rate of growth of the muscles of mastication, including the tongue, and with the development of the teeth. It is therefore not to be expected that the base of the skull and face should grow at the same rate as the vault and orbit. The base of the skull is formed

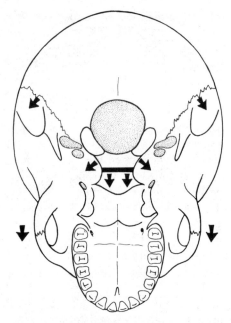

Fig. 31. Growth of base of skull. The main points at which growth in length and breadth take place are indicated by arrows.

in cartilage, and growth may continue here, at the joints between the ethmoid, the sphenoid, and the occipital bones [Fig. 31], until perhaps the twentyfifth year, when the cartilaginous growing plate between the occipital and sphenoid bones is obliterated. This plate is the major site of growth in length of the skull, and growth here pushes the mandible forwards [Fig. 29]. The face also grows for a similar time, the long period being related to the development of the jaws and the nasopharynx. It is for this reason that you do not acquire your permanent facial appearance until about the age of 25.

As the teeth develop, so the upper jaw is increased in size by surface apposition; bone is removed from its inner aspects to keep its proportions constant, and, concurrent with this, the bones of the face are excavated by osteoclastic activity to form the air sinuses which lighten the front of the skull. The process is essentially similar to the growth of the inside of the vault of the skull, but the sinuses are filled with nothing but air. The largest of them is the maxillary sinus [FIG. 29], which grows in spurts following the times of eruption of the maxillary teeth related to it. The frontal sinus above the orbit excavates the frontal bone of the forehead, which is at the same time added to by surface deposition to form the brow ridges. The increase in the length of the skull which occurs at the time of adolescence is largely due to this cause.

The mandible is very small at birth, and consists of two separate halves joined by fibrous tissue in the midline. The two halves fuse during the first year of life, and the main subsequent growth of the mandible is in its length, although, since the mandible articulates with the skull, its width must increase to correspond with the width of the skull. The 'growing point' of the mandible at this stage is in the head of the bone, which lays down new bone by surface apposition, so pushing the mandible downwards and forwards. At the same time there is absorption of bone in front of the ramus and addition of bone behind it, so that more room is provided for the eruption of the teeth [FIG. 32]. One consequence of this is that the angle of the mandible, where the ramus meets the body, becomes progressively reduced with increasing age from about 140 degrees in the infant to about 120 degrees in the adult [FIG. 29]. This is part of the general increase

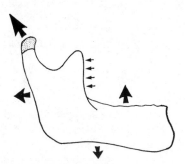

FIG. 32. Growth of mandible. Growth at the head of the mandible pushes the chin downwards and forwards. At the same time, appositional growth in the regions shown increases the vertical and horizontal dimension of the bone, while appositional growth on the outer surface increases its breadth. During this deposition, absorption occurs on the front of the ramus as shown by the small arrows, and on the inner surface of the bone to maintain the relative thickness constant.

in 'uprightness' of the face, much of which takes place round about adolescence.

The maximal growth activity in the region of the face at the time of the adolescent spurt is found in the mandible, which has lagged somewhat behind the rest of the face. Most of the growth at this time occurs in the ramus, but there is also a considerable increase in the length of the body and in the vertical height from the incisor teeth to the chin. Growth in the width of the face and jaws is completed relatively early; anteroposterior growth continues longer, and vertical growth longer still.

The vertical diameter of the orbit in the new-born skull is approximately equal to the distance between the lower margin of the orbit and the lower border of the body of the mandible [Figs. 29, 45], and in the adult skull it is much less than this distance; the difference is due to the growth of the mandible and maxilla.

The way in which the mandible grows means that the teeth have to move forwards to make room for the progressive eruption of the molar teeth. The anterior walls of the tooth sockets are therefore continuously eroded, while bone is added corres-

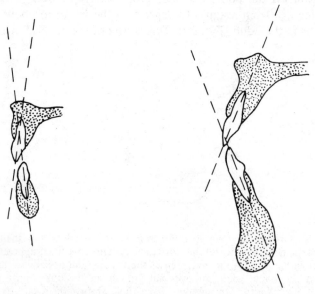

Fig. 33. Alteration with growth of the angle made by the incisor teeth of the upper and lower jaws.

pondingly to their posterior margins. The deciduous incisors in the maxilla and mandible meet each other almost vertically, but their permanent successors slant forwards to meet at an obtuse angle [FIG. 33].

If the maxilla and mandible do not grow at the same rate, the eruption of the teeth may be interfered with, and various disabilities may follow the failure of the teeth to meet each other properly. The way in which the upper and lower teeth fit together (the 'occlusion') has a profound effect on the later stages of growth of the jaws.

The mastoid process behind the ear does not begin to develop until after birth, and the associated bone forming the passage down to the ear drum (the external auditory meatus) is also deficient in the infant, and has to 'catch up' with the growth of the remainder of the bones on the side of the skull. It has been suggested that the growth of this region is stimulated by the pull of the sternomastoid muscle when the baby begins to raise up its head [p. 132], but this is largely speculation, and the relatively minor influence of muscle pull on the shape of long bones [p. 66] lends it little support.

Vertebral Column

Increase in length of the column is determined by the growth of its individual parts. Different sections of the column grow at different rates, and the lumbar and sacral vertebrae, which at birth are relatively smaller in relation to their adult size than are the thoracic and cervical vertebrae, have to grow most. Individual vertebrae grow in thickness by deposition of bone in the cartilage above and below the vertebral bodies; annular epiphyses appear in childhood [FIG. 34] and do not close until well after most of the other bones of the body have ceased to grow, around the age of 25. The epiphyses in the upper thoracic region stay open longer than those elsewhere in the column.

Much less is known about the growth of the intervertebral discs, which account for one quarter to one-third of the length of the vertebral column, and thus make a major contribution to the height of the individual. This contribution is variable, since the discs are compressible, and are therefore responsible for some of the diurnal variation in height [p. 23]. Radiological investigation has shown that in children the disc below the eighth thoracic

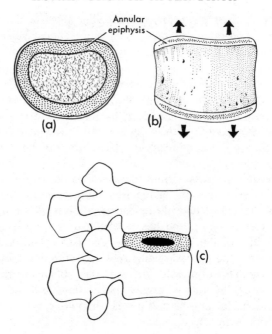

FIG. 34. Growth in height of vertebral column

(a) Body of vertebra seen from above, showing annular epiphysis.
(b) Body of vertebra seen from in front, showing epiphyses.
(c) Two lumbar vertebrae viewed from the side, showing the construction of the fibrocartilaginous disc between their bodies. The pulpy central portion of the disc is shown in black, surrounded by the fibrous part of the disc.

vertebra, after a period of growth around the time of birth, does not increase significantly in vertical diameter between the ages of 6 months and 8 years. On the other hand, the thickness of the disc below the fourth lumbar vertebra increases steadily up to the age of 2 years, and more rapidly thereafter.

The notochordal cells which form the fetal nucleus pulposus begin to degenerate and die even before birth, so that by the age of about 10 years no cells are left and the nucleus is composed of white semifluid material. This is then gradually and progressively invaded by fibrous tissue from the anulus fibrosus, so that the boundary zone between the two components of the disc becomes blurred.

NERVOUS SYSTEM

In the new-born the brain accounts for 10–12 per cent of the body weight, and it doubles its birth weight in the first year of life. By the age of 5 or 6 it has tripled its birth weight, but after this growth slows up rapidly and progressively, so that the adult brain is only about 2 per cent of the body weight.

All the major surface features of the cerebral hemispheres are present at birth, but the cerebral cortex is only about half its adult thickness, and it is much more primitive in microscopic appearance. The spinal cord is about 15–18 cm long at birth, and its lower end is opposite either the second or the third lumbar vertebra—about one vertebra lower than in the adult, in whom it reaches a length of approximately 45 cm. This is because the spinal cord does not grow quite so much as the vertebral canal which contains it, and therefore appears to rise up in the canal.

The nerve fibres in the brain and spinal cord at birth are far from completely mature, since many of those which will ulti- mately receive a coating of myelin [FIG. 21] have not yet done so. In general, the main sensory pathways receive their myelin sheaths before the motor pathways, and at birth all the major sensory tracts are fairly well advanced, while the motor pathways are only just beginning to myelinate. However, the local reflexes related to swallowing and sucking appear before birth, and the cranial nerves which subserve them are well advanced. The optic pathways are well myelinated except for the optic nerves them- selves, but the auditory pathways do not become myelinated completely until about 2 years of age. Myelination of the cerebral cortex is said to continue into adult life.

While it is probably not true, as was formerly thought, that functional development runs directly parallel with progress of myelination, it is certain that not all the tracts and pathways in the nervous system are fully functional at birth.

In the skin there are marked changes in the neural pattern. During the third month of intra-uterine life the epidermis begins to stratify, and it is immediately invaded by branches from the cutaneous plexus of nerves. But before birth, as the organized endings appear in the skin, the intra-epidermal fibres and endings withdraw, so that after birth only a few remain, though their number may increase in certain skin diseases. In an infant the organized sensory endings, such as the Meissner corpuscles, are closely packed together, but as the skin grows they become thinned out, much as spots painted on a balloon separate from each other as the balloon is inflated. There are therefore many fewer such endorgans per unit area in the adult than in the child. For different reasons [p. 214] there are fewer still in old people. It is not known whether a similar thinning out by separation occurs in the population of proprioceptive endorgans such as muscle spindles, or whether the network of 'free' endings in the skin and elsewhere develops a wider mesh with growth.

The eye grows according to the neural pattern and reaches adult size by the age of 14. The eyeball is at first too short for its lens, so that most infants have about 1 dioptre of long-sightedness. As it grows, the eyeball becomes about 2 mm longer, but about 60 per cent of the alteration in focussing this enlargement would cause is eliminated by a simultaneous reduction in converging power of the cornea and lens, so that the difference caused by growth just about cancels out the refractive error of the baby. The lens is peculiar in that it continues to grow throughout life, becoming about one-third bigger at the age of 60 than it is at 20. The outer ear grows according to the general body pattern, but the inner ear, the middle ear cavity, and the drum are of almost adult size at birth and grow very little.

In the infant the arachnoid granulations are small and few, but as the nervous system grows they become larger and more numerous, and bony excavations are formed in relation to them on the inside of the parietal bones alongside the superior sagittal sinus. These pits are radiologically prominent in old age, by which time the granulations may have reached a diameter of a few millimetres.

CARDIOVASCULAR SYSTEM

Immediately after birth a considerable reconstruction of the cardiovascular system has to take place. In the fetal stage of existence the blood is oxygenated in the placenta: there would be no point in driving it round the lungs of the fetus, which at this stage are functionless. The blood which arrives from the placenta and travels along the umbilical vein through the ductus venosus in the liver to the inferior vena cava and so to the right atrium of the heart is therefore largely shunted through a hole in the interatrial septum (the foramen ovale) into the left atrium, so by-passing the pulmonary circulation [FIG. 35]. From the left ventricle it is circulated round the body, returning to the heart by means of the two venae cavae. The deoxygenated blood from the upper part of the body, travelling in the superior vena cava, mostly passes through the right atrium into the right ventricle and so into the pulmonary artery, whence the greater part of it rejoins

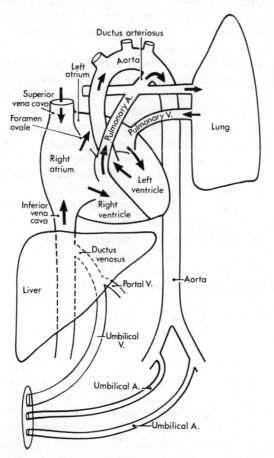

FIG. 35. Fetal circulation. The diagram omits everything except those features essential to an understanding of the main path of the blood— e.g. only one pulmonary vein is shown (see text for description).

the aorta through the ductus arteriosus, again by-passing the lungs. Since this inflow of venous blood to the arterial side of the circulation occurs further from the heart than the origin of the main arteries to the head and neck and the upper limbs, it follows that these regions obtain a better oxygenated supply of blood than do the abdomen and lower limbs. (The blood from the lower part of the body mixes with the arterial blood in the inferior vena cava, but otherwise there is only a limited mixing of oxygenated and deoxygenated blood in the right atrium.) The small and inactive lungs also receive a very poor blood supply. Finally, the blood in the arterial tree is returned along the umbilical arteries to the placenta for reoxygenation.

At birth blood ceases to flow to and from the placenta, and the umbilical vein, the ductus venosus, and the umbilical arteries are immediately obliterated by muscular contraction of their walls and by clotting of the blood within them. The vessels and their contents are eventually converted into fibrous cords by a process which will be described in relation to the healing of wounds [p. 170], and are then known respectively as the ligamentum teres, the ligamentum venosum, and the obliterated umbilical arteries.

The sudden shift of the site of oxygenation from the placenta to the lungs with the taking of the first breath means that the persistence of the pulmonary by-passes in the circulation would result in the blood not being properly oxygenated, and therefore the foramen ovale and the ductus arteriosus have to be closed. The closure of the foramen ovale takes place in two stages. There is first a functional closure, which depends on pressure changes in the atria of the heart consequent on the flow of blood through the newly expanded lungs, and later an anatomical closure, in which the flaps of the valvular opening adhere together and fuse. Frequently the anatomical phase of closure is incomplete, and a hole persists between the two atria. However, so long as the hole is small and the pressure on either side of it remains more or less equal, there is little functional disturbance. On the other hand, should a substantial leakage occur from the right to the left side of the circulation, either between the atria or elsewhere, the blood circulated to the body may be incompletely oxygenated. In such cases there is often a failure of general growth to progress properly [p. 152].

The ductus arteriosus is first shut off by a sustained muscular contraction of its walls, and later, like the umbilical vessels, becomes converted into a fibrous cord, the ligamentum arteriosum. If the growth processes involved in closure of the ductus fail to obliterate the channel, blood may pass from the aorta (high pressure) into the pulmonary artery (low pressure) and the resulting disability may require treatment by surgical means.

The heart is relatively bigger at birth than it is in the adult. In the new-born it weighs about 0·75 per cent of the body weight, whereas in the adult it accounts for about 0·33 per cent. This change is readily seen in radiographs of the chest, for in the infant the heart occupies about 40 per cent of the lung fields, and in the adult only about 30 per cent. The heart of the infant lies more or less transversely, and is later pushed downwards by the expanding lungs, so that in the adult it comes to lie more obliquely, and also lower in relation to the rib cage. The heart may continue to increase in size and weight during adult life, but much of the increase is due to the deposition of fat.

At birth both ventricles have walls of more or less the same thickness, but those of the right ventricle, in conformity with its less arduous task, grow less than those of the left ventricle, so that in the adult they are only about half as thick. There is quite a marked adolescent spurt in the heart muscle, and this is reflected in a rise in blood pressure at adolescence, together with a slowing of the heart rate.

During normal growth there is little or no increase in the number of cardiac muscle fibres, but the cross-sectional area of each fibre increases about seven times to the average adult value of roughly 45 micrometres. As this happens, the number of small vessels increases, so that the number of muscle fibres supplied by each such vessel falls from about six in the new-born to approximately one in the adult. But because the whole system is increasing in size, the number of vessels per unit area of tissue remains constant throughout. It is important that if the fibres in the heart wall are made to enlarge still further [p. 198], there is no corresponding increase in the number of vessels.

As regards the conducting system, there is a marked increase at birth in the amount of collagenous material among the fibres of the sinu-atrial node, and this is progressive with age. The nodal muscle fibres, unlike those of the atria, do not increase in

PLATE 1. *Examples of different types of cells*

All the photographs are at the same magnification: the secondary oocyte in (*a*) is approximately 100 micrometres (one-tenth of a millimetre) in diameter, and may be used as a measuring rod by which the size of the other cells can be appreciated. (Original photomicrographs by kind permission of Dr. N. Thomas.)

(*a*) A secondary oocyte in the ovary. This spherical cell is the precursor of the mature ovum, which is the largest cell in the body. The section has passed to one side of the nucleus, but shows well the contrast in size between the oocyte and the small cells, with darkly staining nuclei, which surround the oocyte and form a capsule for it.

(*b*) Multipolar nerve cells. The cell bodies give rise to long thin processes by which the cells are put in communication with other nerve cells, sensory endorgans, or muscle fibres. These processes may be as much as a metre or more in length (the distance from the lower end of the spinal cord to a muscle in the sole of the foot in a large adult). The preparation has been impregnated with silver in order to show up the cell processes, and this necessarily obscures the details of cell structure; the nuclei, for example, are not visible.

(*c*) Fibroblasts from a healing wound. These undistinguished-looking cells, whose outlines are difficult to make out, are associated in some way with the production of the fibre content and much of the matrix of the tissues. They are of great importance in the process of wound healing. The nuclei are clearly outlined, but the cytoplasm is indistinct; the cells are surrounded by a hazy background of collagen.

(*d*) Fat cells. The cytoplasm of the cells is almost completely replaced by a vast accumulation of fat, which pushes the dark-staining nuclei to one side of the cell. This fat is removed during the process of preparing the specimen, and the sites which it occupied in life thus show up as clear empty spaces. Notice the shapes taken up by the cells as they become closely packed together.

(*e*) Somatic muscle fibres. The vertically-running fibres show marked cross-striation, and the nuclei have been pushed to one side, out of the way of the contracting elements. Each muscle fibre contains many nuclei, and an individual fibre may reach a length of up to 30 centimetres.

(*f*) Cartilage cells. The cell nuclei show up as dark spots lying in cavities in the relatively featureless matrix of hyaline cartilage. The cytoplasm has shrunk away from the boundaries of these cavities during preparation of the specimen, but can be seen surrounding the nuclei of several of the cells.

(*g*) Osteoblasts. The small irregularly shaped osteoblasts, with rounded dense nuclei, form a layer on the surface of developing bone (upper left), to which they are adding new material.

(*h*) Osteoclasts. Two of these large multinucleated cells are shown alongside a portion of bone which is undergoing absorption.

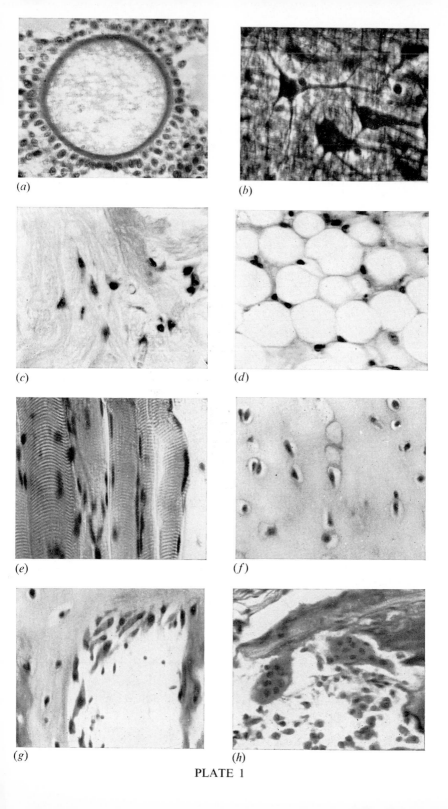

PLATE 1

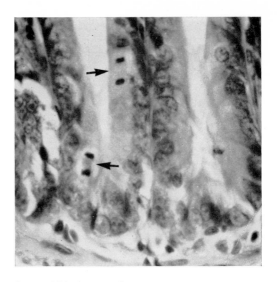

PLATE 2(a). *Mitotic figures*
Two mitotic figures (arrowed) in the wall of the intestine. (Original photomicrograph by kind permission of Dr. N. Thomas.)

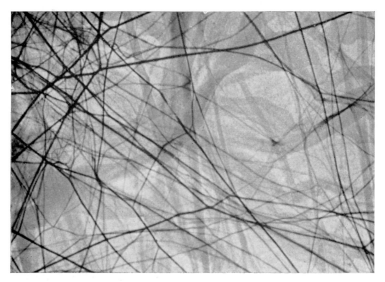

PLATE 2(b). *Collagen and elastic fibres*
A spread of mesentery showing collagen fibres (thick, pale bundles which do not branch) and elastic fibres (thin, branching, darkly stained). (Original photomicrograph by kind permission of Dr. N. Thomas.)

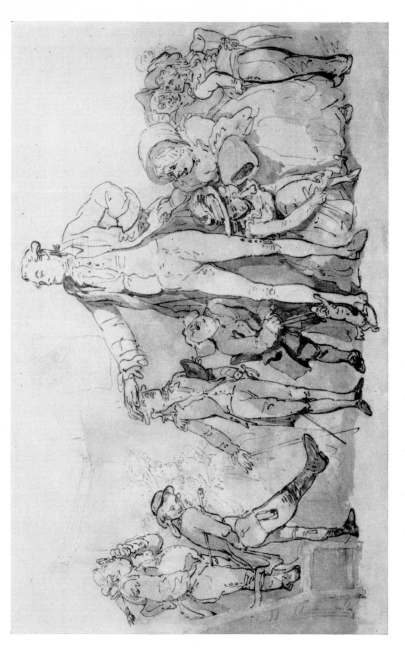

PLATE 3. *Patrick Byrne, the Irish giant* A caricature by Thomas Rowlandson.

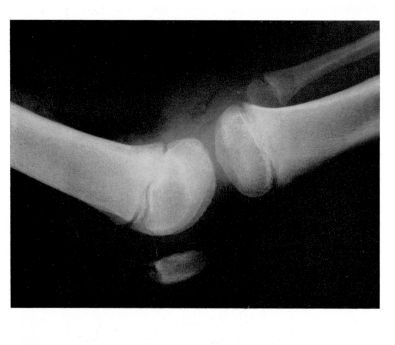

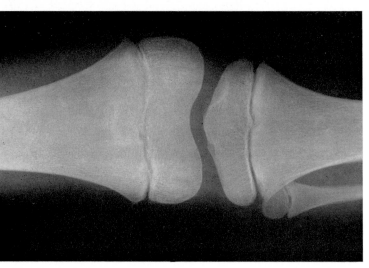

PLATE 4. *Epiphyses*

Radiographs of the knee of a 10-year-old boy showing the epiphyses at the lower end of the femur and the upper ends of the tibia and fibula. Note the clear epiphyseal lines corresponding to the epiphyseal plates of cartilage. The patella is well ossified.

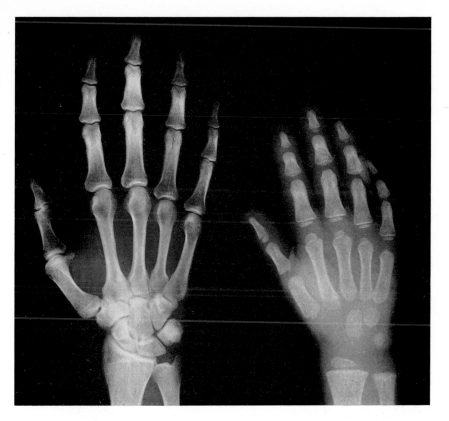

PLATE 5. *Skeletal maturation*

Radiograph of the hand of a 25-year-old man compared with that of a boy of 3½ years. In the child's hand the secondary centre for the lower end of the radius has appeared, but not that for the ulna. There are primary centres for only four of the carpal bones, the capitate (the largest), the hamate, the triquetrum, and (just appearing) the lunate. Secondary centres for the metacarpals and proximal phalanges are present and fairly sizable, except for those of the thumb, and several, but not all, of the secondary centres for the other phalanges are appearing. In this case skeletal and chronological ages show no divergence.

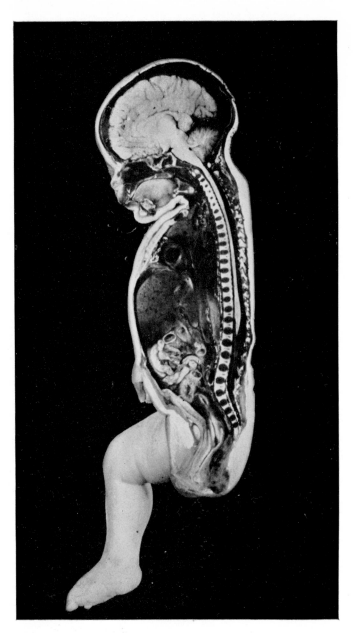

PLATE 6.
Midline section through stillborn baby. There are only two spinal curvatures, both concave forwards. The pelvis is very small, the liver large, the abdomen protuberant, the heart high, the neck short, and the tongue horizontal and wholly within the mouth. (Photograph by kind permission of Emeritus Professor R. Walmsley)

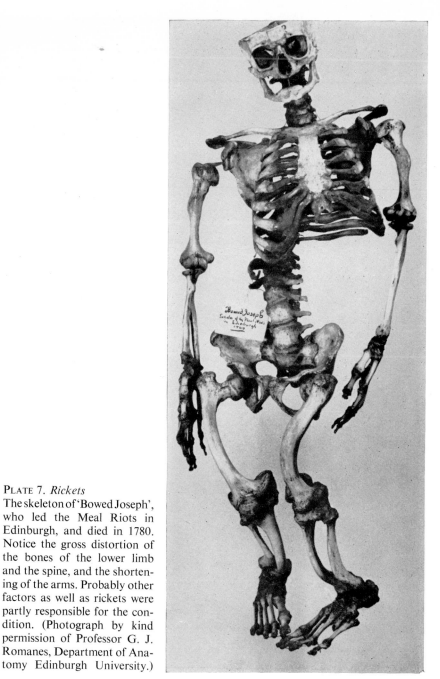

PLATE 7. *Rickets*
The skeleton of 'Bowed Joseph', who led the Meal Riots in Edinburgh, and died in 1780. Notice the gross distortion of the bones of the lower limb and the spine, and the shortening of the arms. Probably other factors as well as rickets were partly responsible for the condition. (Photograph by kind permission of Professor G. J. Romanes, Department of Anatomy Edinburgh University.)

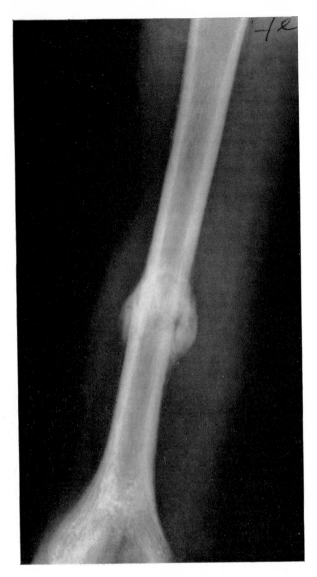

PLATE 8.
Radiograph of middle of shaft of fractured humerus (in good position) showing the formation of provisional callus. The subsequent course of events would be roughly as indicated in FIGURE 59.

diameter very much, and it has been suggested that this is because the atrial fibres have to perform against an increasing load, while the nodal fibres simply continue to initiate impulses and do not undergo mechanical stresses.

The arteries and veins grow in parallel with the growth in weight and height of the body; as they are called on to bear an increased pressure, so their walls thicken [p. 198] in response. Little is known about how blood vessels elongate to keep pace with body dimensions. Indeed, certain arteries grow longer than the distance which has to be travelled, and so become tortuous. In some cases the reason for this is apparent; for example, the lingual artery supplying the tongue is tortuous because it has to be able to stretch out to allow the tongue to be fully protruded. But in others, such as the splenic artery, it is not clear why the artery has to be tortuous or what stimulus to its growth brings it into this condition.

The number of valves in a given vein seems to remain constant during postnatal growth, and the evidence suggests that all the venous valves have appeared by the sixth month of fetal life.

In experimental animals arteriovenous anastomoses can be formed in response to increased local circulatory demand, and may disappear again when the stimulus is removed. Whether this occurs in man is not known. Another response to the same stimulus is an increase in the density of the capillary network of the tissue concerned.

HAEMOPOIETIC SYSTEM

At birth the red bone marrow occupies most of the cancellous spaces and medullary cavities of the skeleton, but during childhood much of it is replaced by fat. By puberty none remains in the long bones except in the upper ends of the shafts of the femur and humerus, and in the normal adult red marrow is found only in the vertebrae and in the flat bones such as the sternum, the ribs, the innominate bones and the scapulae.

There is a steady increase in the blood volume during childhood and also in the number of circulating red cells per unit volume. The red cell count is relatively high at birth, and falls, together with the haemoglobin content of the blood, until about the third or fourth month after birth, when both figures begin to increase again. During the adolescent spurt there is a considerable increase in both haemoglobin and the numbers of red cells, and at this time the red count in boys rises above that for girls, and remains higher for the rest of life. At the same time boys develop a considerable increase in the alkali reserve of the blood. These changes are physiologically necessary for the efficient use of the increased muscular development in the male, and the obvious conclusion is that they are mediated by some hormonal influence, but little is yet known about details.

The white cell count at birth is about 20 000 per mm^3, and falls to about 12 000 by the end of the first week, after which it gradually approaches the adult range; there is no adolescent spurt in the number of white cells.

LYMPHATIC SYSTEM

Lymphoid tissue has a distinctive growth curve [FIG. 27], reaching its maximum development before puberty, and thereafter declining in both relative and absolute size, probably under the influence of the hormones produced by the gonads. The thymus gland is a prominent feature of the chest of an infant, extending from the neck down into the anterior mediastinum and occupying much of the superior mediastinum. At this time of life it weighs anything from 2–17 g, with an average of 13 g. It is intensely active, and the lymphocytes which it produces are exported via the blood stream to the lymph nodes and other centres of lymphocytic activity. There they appear to settle down to form daughter colonies which later become self-supporting. This is part of the mechanism by which the body establishes its system of immunity. In the adult, the thymus may be difficult to find, and in an old person it may be represented largely by a scanty mass of fibrous tissue. It reaches its maximum weight of approximately 35 g about the age of 12 in girls and 14 in boys, and rapidly regresses thereafter.

After birth the lymph nodes increase in size more than in number; the average total weight of the mesenteric nodes is said to increase nearly twenty times, but their number only about three times.

The tonsils and adenoids, so massive in the primary school child, usually settle down, unless they are diseased, to a very inconspicuous existence in the adult. Their maximum size is usually attained about the age of 6 years. The Peyer's patches of the ileum also become much less obvious, and in old age may virtually disappear. The lymph nodes do not regress so dramatically, and may remain prominent—especially in the inguinal region—for many years of adult life. The spleen, which is often considered as part of the lymphoid system, grows with the general body growth curve rather than with the other lymphoid tissues, though the lymphoid tissue in it decreases along with the rest of such tissue elsewhere.

Lymph channels are readily and speedily repaired [p. 178], and their growth in such circumstances can be studied in a transparent chamber in the ear of a rabbit; it appears to be virtually identical with the corresponding processes in blood vessels.

RESPIRATORY SYSTEM

In contrast to the nervous system, the respiratory system is relatively very small at birth, but the lungs are not completely unexpanded, since some respiratory movements take place in the uterus. These partly fill the lungs with amniotic fluid, which is very rapidly absorbed after birth. After the first breath the lungs begin to grow very fast; they increase three-fold in weight and six-fold in volume during the first year. At the same time changes occur in the air passages. The terminal parts of the conducting tubes give rise to sprouts forming additional alveoli, and themselves become modified into respiratory bronchioles, while the existing respiratory bronchioles become modified into alveolar ducts. In this way the available respiratory surface area is increased, and at the same time the originally small conducting tubes rapidly grow in diameter. Altogether about six generations of conducting tubes are added to the bronchial tree after birth, most of the additions taking place in the first few months. In the later stages of growth the main feature is an enlargement of existing alveoli rather than further development of new ones.

Before birth the distal portions of the tree and the alveoli are lined by cuboidal epithelium, which flattens as soon as the lung is distended and respiration begins.

The respiratory system, after the first few weeks, follows the general growth curve, but, because the respiratory passages grow faster than the thoracic and cervical portions of the vertebral column [p. 79], the larynx and trachea descend relative to it, so that the bifurcation of the trachea, which in the infant is opposite the third thoracic vertebra, comes to lie opposite the fourth by about the seventh or eighth year. The cricoid cartilage, opposite the fourth cervical vertebra in the infant, descends to the sixth cervical vertebra by the same time. The hyoid bone, which is level with the mandible before the onset of the adolescent spurt, drops rapidly at puberty to a somewhat lower level. The infant diaphragm, which is forced up by the pressure of the contents of the abdomen, descends to its adult position as the pelvis enlarges and the abdominal viscera descend [p. 119].

The larynx grows slowly until puberty, when its growth speeds up in both sexes, but markedly more so in boys, the vocal folds doubling in length from 8 mm to 16 mm within a year; this increase in its size is associated with the 'breaking' of the voice. Further growth subsequently takes place to the adult length of about 25 mm in the male and 20 mm in the female.

DIGESTIVE SYSTEM

The development and eruption of two successive sets of teeth in a well-charted sequence affords a means of assessing the maturity of the child, and the process is considered in Chapter 6. The tip of the tongue in an infant is usually stubby and rounded, and the mobile and more pointed tip of the adult tongue develops later, with chewing and with speech.

At birth the tongue lies wholly in the mouth, but during the first four or five years of life its posterior part descends with the larynx and comes to form part of the front wall of the pharynx. The new-born baby is a nose breather, and until the tongue descends the high position of the larynx and tongue allows him to continue to breathe freely while fluid is passing down on either side of the epiglottis and the uvula, which are in contact during swallowing.

The oesophagus has a diameter of about 5 mm in the new-born, and both its upper and its lower ends are about two vertebrae higher than in the adult.

In the new-born the stomach is very small, having a capacity of approximately 30 ml, and lies horizontally. With the beginning of feeding its capacity rapidly enlarges, tripling during the first two weeks. By the end of the first year it is capable of holding about 500 ml, and in the adult its capacity often reaches 1500 ml. Like the respiratory system, the digestive system 'descends' with growth; the long axis of the stomach becomes more oblique, and the length of the small intestine is doubled by the time of puberty. The intestinal tract at birth is very thin-walled because the musculature is not yet well developed; the mucosa and submucosa are more robust. Villi continue to form in the small intestine up to the time of puberty.

The posterior wall of the abdomen is relatively shallow in the infant; the lumbar curvature does not yet exist, and the vertebral column does not project so far forwards as it does in the adult.

Because the infant pelvis is small [p. 119], very little of the small intestine can be accommodated in it; in the adult there is more room, and the small intestine, and indeed even the stomach, may extend into the pelvis. The caecum in the fetus is conical, and the appendix is related to the apex of the cone. Differential growth later causes the caecum to descend, and alters its shape to a rounded cup; the appendix moves round to its inner side. The ileocaecal peritoneal recesses are usually more distinct in children than in adults.

The digestive organ which contributes most to the weight of the body is the liver, and at birth the liver is relatively extremely large, its lower margin being palpable below the costal margin. At this stage, it weighs 5 per cent of the body weight, and this declines to about 2·5 per cent in the adult. Its size at birth reflects the importance of the many functions, including blood formation, which it is called upon to perform during intra-uterine life.

URINARY SYSTEM

In general the growth of the kidneys corresponds to the growth of the body as a whole. At birth the two kidneys weigh about 23 g; this weight doubles in six months and trebles by the end of the first year. Like the brain, the kidneys are not fully functional at birth, and many of the renal tubules are not yet formed; the glomeruli enlarge considerably after birth, and in the new-born the cortex is much thinner relative to the medulla than it is in the adult. Probably no new glomeruli appear after the first few months of postnatal life. The kidneys of the new-born exhibit lobulation, which reflects their development in the embryo, but with growth this lobulation usually disappears, though traces of it are often still found in the adult.

An important histological feature of the new-born kidney is the fact that the visceral layer of the capsule of the glomeruli is cubical rather than flattened. It is not until after the first year of life that this cubical epithelium is fully replaced by thin pavement epithelium, and during this year filtration by the kidney is relatively poor.

The growth of the kidney seems to depend on the physiological

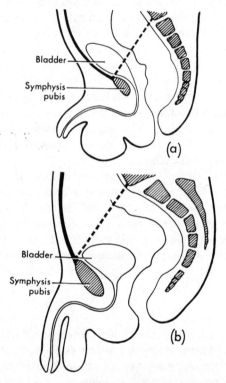

Bladder

Symphysis pubis

(a)

Bladder

Symphysis pubis

(b)

FIG. 36. Position of urinary bladder. Diagrammatic section through male pelvis.

(a) Infant. The bladder is almost entirely an abdominal organ.
(b) Adult. As the pelvis expands with growth, the bladder sinks into it, so that when empty it does not extend into the abdominal cavity.

work it has to do; if one kidney is removed from an experimental animal the remaining kidney can be induced, by giving a high protein diet, to become larger than the two kidneys combined would have become [see also pp. 178, 200].

The urinary bladder is an abdominal organ in infancy, but descends into the pelvis as the latter becomes more capacious [FIG. 36]. It follows from this that the infant ureter is relatively much shorter than in the adult, and that it has no pelvic portion. It also follows that the shape of the bladder alters with age. In the infant it is somewhat cigar-shaped, and it does not achieve its adult pyramidal shape until about the sixth year, by which time it has come to occupy more or less its adult position. In the infant the posterior surface of the bladder is completely covered by peritoneum.

REPRODUCTIVE SYSTEM

The gonads and external genitalia grow very slowly in childhood until the adolescent spurt [FIG. 27]. During its development the testicle 'descends' down the back wall of the abdomen and through the inguinal canal into the scrotum. By the time of birth it has usually either arrived at its destination, or has at least descended beyond the superficial inguinal ring, but there are considerable variations in the timing of this operation, and frequently the testicle does not descend fully until some years after birth. Since the descent of the testicle is necessary for the full maturation of the spermatozoa, a process which requires a temperature lower than that of the abdominal cavity, retention of the testicle in the abdomen is associated with sterility, and, indeed, the seminiferous tubules of such a testicle may become converted into fibrous tissue. Nevertheless, the endocrine function of an undescended testicle may be normal, and the secondary sexual characteristics [p. 104] may appear as usual.

The seminiferous tubules of a normal testicle are solid until puberty, and the organ remains small and feels soft. At puberty [p. 104] the tubules become canalized and, as spermatogenesis begins, the testicle enlarges and acquires its characteristic resistant feel, due to the pressure of the spermatozoa being formed inside it. The adult testicle is roughly 40 times as heavy as the new-born testicle.

The prostate gland and the rectum are the only major organs in the male pelvis at birth. The prostate grows steadily till the time of the adolescent spurt, when it doubles its size over a period of 6 to 12 months. It continues to grow more slowly for some years after the spurt is completed; the secretory component of the gland increases relative to the musculofibrous tissue. At puberty the mucous membrane lining the seminal vesicles, which until this time has been thin and insignificant, becomes greatly thickened and folded in response to the secretion of testosterone by the testicle.

At birth the penis is relatively large; and the prepuce is often imperfectly separated from the glans; it may be some years before full separation takes place. The male urethra triples its length during post-natal growth, the spongy part growing most.

The ovaries at birth are small, though relatively large in comparison with the testicles. Oogenesis is complete at or shortly after birth, and no new oocytes are formed afterwards. There is little growth of the ovaries until puberty, when the ovarian cycle begins, under the control of the anterior lobe of the pituitary gland [p. 166]. At this time the ovaries increase in size to a maximum of about 20 times the weight of the new-born organs; in old age they regress again, and may be largely replaced by fibrous tissue [p. 216]. Like the bladder, the ovaries are abdominal organs in the infant, and are large in relation to the uterus. They enter the ovarian fossae by about the sixth year, just as the bladder is enabled to descend into the pelvis. Before puberty the surface of the ovary is smooth, but after ovulation has become established it becomes progressively more and more scarred and 'cobbled'. During reproductive life the total number of oocytes becomes depleted from about 750 000 at birth to about 10 000 in women aged 45, although only some 450 ova are produced during that time.

The uterus is relatively large at birth, presumably due to the influence of maternal hormones passing across the placenta. When this stimulus is removed, the uterus loses weight and does not regain its birth weight until the hormonal stimuli of impending puberty activate its growth again. A curious feature is that the cervix of the uterus is considerably larger than the body of the organ until adolescence, when the body starts to grow so rapidly that it outstrips the cervix and comes to be double its length.

Because of the crowded situation in the pelvis, the uterus at birth lies more or less in the same plane as the vagina. As the bladder descends into the pelvis, so the uterus bends forwards into its adult posture of anteversion and anteflexion. The uterine tubes grow along with the uterus, in a similar pattern.

During pregnancy the uterus increases its weight from about 40 g to nearly a kilogram. Most of the increase is due to growth of the muscular wall of the organ, in which the muscle fibres increase in length, in thickness, and in number. But the connective tissue components of the uterine wall also increase, and the endometrial cells enlarge and accumulate glycogen, becoming converted into decidua. The glands in the region of the cervix form a thick mucous plug which blocks the cervical canal.

Two main factors appear to be responsible for these changes. The first is the stimulus provided by the hormones secreted by the corpus luteum of pregnancy and the placenta, and the second is the mechanical stretching of the uterine wall by its contents.

After delivery the uterus contracts and begins the process of involution; by the end of the first week it weighs about 500 g, and by the eighth week it has returned to approximately its resting weight.

In both sexes the breasts may be enlarged in the period immediately after birth, and they often produce a certain amount of secretion. This is sometimes referred to as 'witch's milk', since it was a stigma made much of by some of the witch hunters in the seventeenth century. The phenomenon is due to stimulation by maternal hormones, and the breast activity rapidly subsides. The

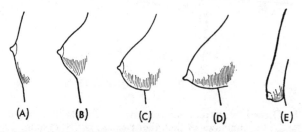

(A)   (B)   (C)   (D)   (E)

FIG. 37. Development of breast. Profiles of the female breast at different ages.

(a) Childhood   (d) Pregnancy
(b) Adolescence   (e) Old age
(c) Maturity

breast then grows along with the general subcutaneous tissue until the time of puberty, when the female breast increases in size and becomes surrounded with secondary sexual fat [FIG. 37]. The nipple enlarges and the areola becomes more pronounced. Thereafter the breast may increase in size slightly with the onset of each menstrual period and subside again after menstruation is completed. During pregnancy it is stimulated to grow by a complicated hormonal mechanism which culminates in the production of milk for the baby.

ENDOCRINE SYSTEM

The pituitary gland is relatively large at birth, weighing about 100 mg, and grows, like the thyroid gland, at a fairly steady rate [FIG. 38]. However, the curve of growth is different in the two sexes, for the weight of the anterior lobe of the female pituitary is greater than that of the male; this superiority becomes evident fairly early in childhood, and is emphasized at the time of the adolescent spurt.

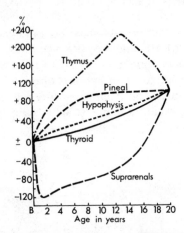

FIG. 38. Growth of certain internal organs. The lower three are endocrine glands; the functions of the other two are not so certain. The curves show percentage of adult weight attained at a given time. (From Scammon, R. E., The measurement of the body in childhood, in Harris, Jackson, Peterson and Scammon (1930) *The Measurement of Man.* University Press, Minneapolis, University of Minnesota, by kind permission of the publishers.)

The parathyroid glands are relatively simple structures, with apparently only one type of secreting cell, until the age of five or six years when a second type of cell begins to appear; it may be that this second type is a modification of the first, and the functional significance of the change is so far unknown.

The suprarenal glands have a growth curve similar to that of the uterus [p. 94], but for a different reason. At birth they still have a thick cortex containing the characteristic fetal or transitional zone, but within a few weeks this begins to disappear, and the three layers of the much thinner permanent cortex become established. As a result the glands become smaller, and do not regain their birth weight until puberty [FIG. 38]. In spite of a well marked adolescent spurt, the adult suprarenals are only about double the size of the new-born organs.

The prominent chromaffin tissue of the new-born regresses during growth. Thus the para-aortic bodies, which are maximal in size during the first three years and measure about 1 cm long, have completely dispersed by puberty, their function being taken over by the suprarenal medulla.

Histological appearances indicate that the pineal body, like the thymus gland [p. 88], is most active in late fetal life and early childhood. In the adult it often becomes calcified, so affording the radiologist a useful intracranial landmark.

5

Indices of Maturity

'... *young boys and girls*
Are level now with men'
SHAKESPEARE

RADIOLOGICAL AGE

It is obviously of considerable importance to be able to assess how far an individual child has progressed towards maturity. The chronological age of the child is a most unreliable guide, since children mature at very different rates, and measurements of height and weight do not take us very far. Several methods have therefore been developed to help us to find out whether a child is lagging behind his fellows in the journey towards adult status, or whether he has raced ahead of them.

Perhaps the most important of these methods is the assessment of 'skeletal', 'anatomical', or 'radiological' age (the names are synonymous). During growth every bone goes through a series of changes [CHAPTER 3] which can be recorded radiographically. The times of appearance of primary and secondary centres of ossification can be observed because the calcium content of the centres renders them radio-opaque, and the progressive enlargement of the ossified portion of an epiphysis or of a short bone can be followed in detail.

The sequence of these changes is roughly similar for a given bone in every person, but their timing may differ quite widely according as the local skeletal development is advanced, average, or delayed. As would be expected, there are exceptions to uniformity, and occasionally some of the centres may appear in an unusual sequence, but in general the course of events is reproducible enough to allow of comparisons between different individuals.

Radiological examination therefore allows us to say how far the skeleton of a child has progressed towards the adult condition. It is clearly inconvenient, and potentially dangerous, to X-ray the whole body, and a sample of the general status of the skeleton is all that is generally taken. The wrist and hand are most commonly used for this purpose, because in this region there is a large number of centres of ossification. It is not dangerous to the patient to use the hand, and data on the hand have been available since the earliest days of radiology. Nevertheless, the use of the hand as an index of skeletal maturation carries with it the assumption that the rest of the skeleton will behave similarly, and this assumption is not always true. Accordingly, in cases where there is a severe deviation from normal standards, other parts of the body may be used as a check [p. 101].

At first it was thought sufficient merely to record the presence or absence of the centres in the hand and wrist [PLATE 5]. But it was found that, although satisfactory differentiation between stages of maturation could be achieved in the early part of childhood, at the time of puberty so few centres remained to appear that there was poor differentiation round about this age. Accordingly, the next method involved the recognition of different qualitative steps in the growth of the various centres in the hand region; for a time measurements of the sizes of the centres were also made. Eventually it was realized that every centre of ossification passed through a number of morphological stages which have become known as 'maturity determinators'. These changes in the outline of the centres can be identified with considerable certainty, and are now listed in key diagrams.

On this basis atlases showing skeletal development at different ages were constructed, and it is of some importance to realize how these atlases took shape. First of all large numbers of radiographs of children of a given chronological age were examined, and every individual bone was allotted a median maturity rating for that age. An attempt was then made to find two single films (one male, one female) in which *every* bone would be as nearly as possible at the maturity rating already determined. These films were then accepted as the standard for that particular chronological age.

Such atlases have been of great use in child development

studies. There are, however, two reasons why an atlas should not be relied upon too uncritically, quite apart from the fact that children of different socio-economic classes and different races mature at different rates [cf. p. 24]. The first is that there is considerable genetic variation in the order in which centres appear. The fact that a given centre has not appeared by its 'proper' time is therefore not necessarily an indication that skeletal maturation is delayed, and it is necessary to take into account the general appearances of *every* centre when coming to a decision. For example, the times for the hamate and capitate do not correlate well with those for the other centres. The second objection is that arranging photographs in order of skeletal age implies a unit of measurement which may be called a 'skeletal year'. This unit, unlike a chronological year, is by no means constant; maturation does not proceed at a steady rate, and therefore to say that a child is a year behind the norm means very different things at different stages of development. Nor is it easy to fit a given radiograph exactly into position by extrapolating from one stage of development illustrated in the atlas to another.

For these reasons scoring systems were developed, and nowadays there is a tendency to rely on the 'percentage maturation' in the selected region. Eight stages are recognized for each bone, and points are awarded accordingly. The result of scrutiny of a wrist radiograph is now reported as a percentage of complete over-all maturation—i.e. 100 per cent score. The bone score can be converted into a 'bone age' by the use of centile charts for boys and girls similar to those for height and weight [see CHAPTER 2].

The clinical reason for using such methods is to be able to say whether a child who comes for examination is skeletally retarded, 'normal', or advanced, and the next difficulty is immediately obvious. What criterion are we to adopt to distinguish the 'normal' from the 'abnormal'? Is a difference of two years on the atlas sufficient? Should the difference be so many points on the scoring system? This is exactly the same difficulty as faced us when dealing with the growth charts for height and weight [p. 23], and there is no hard and fast answer. Individual variation in maturation of the skeleton, like individual variation in growth, is so considerable that it is quite possible for a child at one stage of development to be widely separated from the median maturity picture without this necessarily implying any deleterious effect on his future. If, however, serial checks at intervals of six months or a year show a consistent or increasing deviation from the median, the matter has to be regarded more seriously. Even so, radiological age is only one factor in a general examination of the patient. If there is any suspicion that the hand may not be reflecting a true indication of general progress, radiographs of other regions may be taken, and a composite report obtained. For example, a case is cited of an undernourished 8-year-old child whose over-all skeletal age was thought to be just over 5 years, but who had some bones at $2\frac{1}{2}$ years development, and others at $6\frac{1}{2}$ years development.

Skeletal maturation usually proceeds roughly parallel with skeletal growth, and of course maturation and growth both come to an end when the epiphyses close. The factors influencing skeletal maturation are similar to those which control growth, and will be discussed in CHAPTER 7. Prominent among them is sex, for at every chronological age up to full maturity the radiological age of girls is in advance of that of boys by a factor of about 20 per cent or more [p. 142].

DENTAL AGE

Another series of convenient landmarks by which the development of the child may be estimated is provided by the times of appearance of the primary and secondary dentitions.

The primary teeth begin to calcify during the fourth to sixth months of fetal life, and at birth some of them are more advanced than others. At about 6 months of age, the mandibular incisors (usually) are the first of the primary teeth to erupt—that is, to appear wholly through the gum. At this stage calcification of the primary teeth is not yet complete, and it does not become so until about 3 years of age. The crowns of some of the permanent molars are by this time fully developed, and their roots are beginning to form.

At 6 years of age the mouth is said to be 'full of teeth', since at this time there are more teeth in the jaw than at any other. The teeth of the primary dentition have not yet begun to fall out, and those of the secondary dentition are more or less fully formed. Shortly after this the primary incisors are shed, and the first permanent molar erupts. A series of complicated changes takes place in the jaws to allow room for the larger secondary teeth to occupy the limited amount of space left by the shedding of the primary teeth, and it is at this time that many of the difficulties caused by crowding of teeth may become apparent.

The radiographic appearances of the developing jaws and teeth are sometimes used to give an accurate estimate of the developmental age of the individual, but most estimates have been based simply on the times of eruption of given teeth, and the simplest method of obtaining the dental age requires nothing more difficult than enumerating the secondary teeth present; recourse to a table gives the dental age.

The times of eruption of the secondary teeth are shown in FIGURE 39, but it must once again be understood that this sequence is far from invariable, and that the diagram gives little indication of the probable course of events in any single person. Nevertheless certain teeth appear at fairly predictable times. One of these is the second permanent molar, which erupts at about 12 years of age. Because of its regularity it was once used as an indication of whether a child was old enough to go to work in a factory under the terms of the Factory Act, and accordingly it was named the 'factory tooth'. In an era when documentation of chronological age was far from complete such a test had its value. The third permanent molar erupts at a very variable time between the ages of 18 and 80, and has no value as a marker; it was called the wisdom tooth on the assumption that it appeared at about the age of discretion.

There does not appear to be any sex difference in the eruption of the primary teeth, but, just as girls are ahead of boys in skeletal maturity [p. 101] so they lead in the secondary dentition, the biggest difference being in the times of eruption of the canines [FIG. 39].

Difficulties in assessment are introduced by the fact that some investigators have taken 'eruption' to mean the time of first appearance of any part of the crown of the tooth through the gum, while others have waited until the whole of the crown has pushed its way through; there is often a considerable interval between the two stages. Nevertheless, dental age is a useful way of estimating maturity, and is frequently used in combination with skeletal age, both in paediatrics and in criminology. Just as

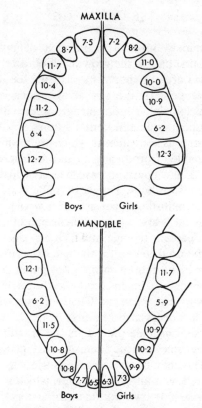

FIG. 39. Mean times of eruption of the permanent teeth. Note that the first to appear is usually the first permanent molar in the lower jaw. The times of appearance of the third molar ('wisdom') teeth are so variable that they have been omitted.

the condition of the skull sutures [p. 73] can be used to obtain a very rough estimate of the age of an adult, so can the condition of the tooth enamel. This, the hardest material in the body, is gradually worn down by a lifetime of eating, and the amount of wear is permanent, since the enamel organ of the tooth dies when it has completed its work, and enamel cannot be renewed. Wear depends not only on age, but on the type of food eaten, and anything like an accurate estimate of age is impossible. Unlike such creatures as the shark, man does not have a reserve dentition, and if a tooth of the secondary dentition is lost, there is no replacement. Such losses are not due to age, but to disease.

SEXUAL AGE

Dental development gives a good idea of progress towards maturity in the first half of the growth period. In the second stage of growth, round about puberty, most of the teeth have erupted, and the development of the sexual apparatus can be used to supplement the somewhat more meagre information afforded at this time by radiographic and dental studies.

Legally, puberty is the time at which the individual becomes functionally capable of producing a child; in England this state is recognized by the law courts to have been reached by the ages of 12 for girls and 14 for boys. The word 'puberty' is often misleading, since sexual maturation consists of a whole series of events spread over several years, with a sequence and timing which vary from person to person; the age at which any one of these events occurs cannot be used to deduce the age of occurrence of any or all of the others. The rate of progress of puberty is independent of the time at which it begins, and the duration of the process varies from about 2 to $4\frac{1}{2}$ years. Children who begin to develop early are thus not necessarily the first to reach sexual maturity.

The earliest sign of male puberty is the growth of the testicles shortly after the time at which the pituitary gland has begun to act on them. This may occur as early as $9\frac{1}{2}$ or as late as 15, and may be measured with an orchidometer, which is simply a series of plastic ovoids of known volume strung together. The ovoid most closely corresponding to the size of the testicle as determined by palpation is taken as giving the volume of the organ. The average adult testicle has a volume of about 20 ml, and a volume of 6 ml may be taken as indicating that puberty has started. At this time mitotic figures abound in the seminiferous tubules, spermatogenesis begins, and testosterone appears in the urine. Following the production of hormones from the testes and suprarenals, the penis enlarges in size, and this is followed by changes in the larynx, the skin, and the distribution of the hair on the body. The timing of the growth spurt of the penis is subject to just as wide variations as the timing for the spurt in height, and at the age of 14 some boys may be practically mature sexually, while others have not even begun their sexual spurt. It is useful clinically to remember that whatever his age, a boy in the early stages of sexual development has usually not begun his maximal rate of growth, and will subsequently add considerably to his height.

The acceleration of the growth of the larynx does not occur until about the termination of the penile spurt, and is presumably due to a direct stimulation of the cells of the laryngeal cartilages and the associated soft parts by testosterone [p. 147]. Pubic hair, on the other hand, may appear before the spurt in height has begun, though it is usually somewhat later; five stages in its development have been distinguished clinically. Axillary and facial hair is usually delayed in appearance for two years or so after the first appearance of the pubic hair. The apocrine sweat glands of the axilla and genital regions are thought to increase in number at puberty, and certainly are aroused to activity at this time. There may be some enlargement of the breasts, and the diameter of the areolae increases. The scrotal skin darkens, and the prostate and seminal vesicles enlarge concurrently with the penis. The first ejaculation of semen usually occurs within a year of the beginning of the enlargement of the penis (Table 4).

In the female the ovaries, like the testicles in the male, increase

TABLE 4

ROUGH TIMETABLE OF SEXUAL MATURATION

	BOYS	GIRLS
Onset	Testicular enlargement begins Seminiferous tubules canalize Primary spermatocytes appear Fine downy straight pubic hair appears	Ovarian enlargement begins Breasts develop to 'bud' stage Fine downy straight pubic hair appears
A year or more later	Secondary spermatocytes present Penile enlargement progressing Pubic hair now coarser and curling	Pigmentation of areolae Pubic hair now coarser and curling
A year or more later	Relative enlargement of larynx beginning First ejaculation	Relative increase of pelvic diameter beginning Menarche; first cycles may not produce ova
A year or more later	Mature spermatozoa present Axillary hair Sweat and sebaceous glands very active	Full reproductivity Axillary hair Sweat and sebaceous glands very active

in size as the first sign of puberty, but this cannot be detected except at operation, although the quantity of oestrogens excreted in the urine gives an indication of ovarian activity. The first external sign of female puberty is the enlargement of the breasts, which precedes the maximal velocity of the adolescent spurt. About the same time the growth of the uterus and vagina accelerates, and the pubic hair makes its appearance. The vaginal epithelium thickens and accumulates glycogen, and the bacterial flora in the vagina changes, with the result that its reaction becomes acid.

Several stages in sexual development have been recognized for the purpose of estimating the sexual age. For example, the maturation of the female breast has been subdivided into five stages, and so has the maturation of the female pubic hair. By far the greatest amount of work has been done in relation to the onset of menstruation (the menarche), which occurs anything up to four years after the first sign of breast development, the average interval being about $2\frac{1}{2}$ years. This is an easily recognizable event, which is well remembered by the subject, and can be timed with some accuracy, although there are difficulties in collecting data. For example, a common method employed is to question women retrospectively regarding their menarche, but it is found that their recollection may differ according to their age at questioning, and that dates tend to be clustered round birthdays, school terms, and so on. A second method is much more accurate, but involves smaller numbers, since it consists in following a group of children longitudinally until the menarche occurs. A third method is statistical, and depends on analysing the percentages of children of accurately known ages who have already started menstruating.

The menarche does not occur until after the peak of the spurt in height is over [FIG. 40], and corresponds to the period of maximal deceleration of growth. A tall girl who has begun to menstruate can be told with confidence that her growth is rapidly slowing down. Many of the early menstrual cycles may not involve ovulation, and full sexual maturity may therefore be delayed for a year or two after menstruation has begun. Conversely, cyclic liberation of ova may occur in some girls much earlier than the onset of menstruation.

The age at which the menarche occurs is found to be more

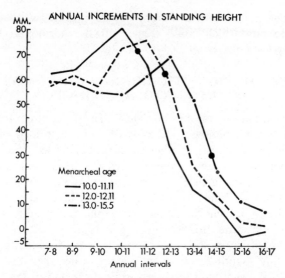

FIG. 40. Time of menarche in relation to adolescent spurt. The three velocity curves for height represent the growth of early, average, and late maturing girls, and the black circles show the average time of menarche for each group. Notice that the menarche occurs after the peak velocities have begun to decline. (Slightly modified from Simmons, K., and Greulich, W. W. (1943) Menarcheal age and the height, weight, and skeletal age of girls aged 7 to 17 years, *J. Pediat.*, 22, 518–48, by kind permission of Professor Greulich and the editor.)

closely related to the radiological age than to the chronological age. Thus, menstruation nowadays begins within a chronological age range of 10 to 16 years, but within a much narrower range of radiological age—from 12 to $14\frac{1}{2}$ years, round about the time of fusion of the epiphyses of the terminal phalanges of the fingers.

Tall girls reach puberty earlier than short ones, but girls with a late adolescent spurt and late puberty are ultimately taller on the average than those who begin menstruating earlier, for they have longer in which to grow [p. 27]. The birth order also has some effect, since puberty occurs earlier if there are several older brothers and sisters; the difference between the firstborn and the sixth child may be as much as a year. It has been suggested that the menarche occurs as a critical weight of about 50 kg (106 lb) is reached. In animals weight is more important than chronological age in determining the onset of puberty, and it is possible that the attainment of a critical weight might involve a change in metabolic rate which could trigger off the hormonal changes necessary for puberty.

ESTIMATION OF CHRONOLOGICAL AGE FROM ANATOMICAL DATA

The determination of chronological age is required for legal purposes when unknown bodies have to be identified or when attempting to establish whether or not a person has reached the age at which the law recognizes responsibility. For example, during the terrorist activities in Palestine in the nineteen-forties it was occasionally found necessary to establish the age of a convicted murderer, who could not or would not produce a birth certificate, in order to determine the legal penalty (death or imprisonment) to be inflicted. More often the problem concerns the age and identification of the victim rather than that of the murderer. A celebrated historical example is that of the 'princes in the Tower', whose skeletons were subjected to anatomical investigations some time ago in order to try to pin-point the date at which they were done to death, and so to establish whether Richard III or Henry VII was responsible.

The wide range of biological variation makes it impossible to say *exactly* how old a person is. A probability can be established by anatomical examination, but nothing more. In adolescence one cannot be accurate to within less than three or four years, and in later life the margin is much wider, since after the epiphyses have closed there is less to go upon, and such indications as the fusion of the skull sutures [p. 73] or the appearance of the pubic articular surfaces [p. 72] have to be used.

NEURAL AGE

We must now turn to a more functional aspect of growth—the progressive development of skilled controlled movements and of the psychological correlates of brain function.

In spite of the large size of the central nervous system at birth, much of it, as has already been pointed out [p. 81], is incompletely functional, and requires a considerable time to develop to the stage at which it can be utilized to the full. A new-born baby has a limited repertoire of purposeful activity. He can cry, suck, swallow, sneeze, move his eyes, defaecate, and micturate. He can taste and smell, but may not be provided with a fully developed sense of pain. He has a remarkably powerful grasp reflex, which he loses in a couple of months. The human baby faces two main difficulties which are not shared by most other animals; he relies on learning rather than inherited behaviour, and he has to develop the skills necessary to hold him upright on a narrow base against the force of gravity and to enable him to use his upper limbs as tools with which to manipulate his environment. It is not surprising, therefore, that motor and sensory development are slow, and that progress can readily be followed in the growing child, and made use of as an index of maturation. A good example is the way in which the apparently random swings of the upper limbs in the baby gradually become refined into purposive handling behaviour: the parts of the limbs nearest the trunk come under control before the more distant parts. A great deal of detailed work has been done on this type of development by Gesell and his colleagues in America, and by others in Britain: it is impossible to give a proper account of it in a book of this size, but certain landmarks of development are shown in TABLE 5 and in FIGURE 41, which will give an idea of the scope of such investigations.

Girls are ahead of boys throughout the phase of motor and sensory development: they are first to acquire the movements necessary for creeping and walking, and first in the development of skills which need fine movements and co-ordination, such as tying bows. They learn to control their bladders earlier than boys.

Girls also mature somewhat earlier in the understanding and use of speech. One of the landmarks commonly recorded by mothers is the age at which their babies say their first word. This occurs at a very variable time, but by about $1\frac{1}{2}$ years of age approximately 30 words are accurately used. By 2 years this number has risen to about 300, and the sex difference has appeared, girls being more advanced than boys. By the age of 3, the child probably knows more than 1000 words, and by 5 more than 2000. The combination of words into language has progressed far enough by the age of 3 for nearly all children to use at least 3 words per sentence; a child of 5 years uses more than 5 words per sentence, but there are very great variations in this as in other phases of psychological development.

TABLE 5

DEVELOPMENTAL LANDMARKS

AGE		
YEARS	MONTHS	
—	2	Follows moving objects with eyes
—	4	Can sit propped up for a short time Moves head to inspect surroundings
—	6	Grasps objects. Begins to bang and shake objects
—	8	May sit unaided
—	10	Creeps: Picks up small objects between finger and thumb One or two words: Tries to help with feeding
1	—	'Cruises' holding on to rail of cot Walks with one hand held: Throws objects on floor Co-operates in dressing: Waves goodbye Puts toy in and out of container
1	6	Walks: runs awkwardly and stiffly Builds towers of 3–4 blocks Can turn pages of a book About 30 words
2	—	Runs without falling: Uses three-word sentences Can turn door knob: Obeys simple instructions Builds towers of 6–7 blocks Bowel and bladder control sometimes good
3	—	Walks erect: can stand on one foot: Climbs Can put on shoes and unbutton some buttons Bowel and bladder control usually established Eats reasonably well by himself Counting begins
4	—	Drawing and copying; printing letters Cleans teeth, washes and dries face and hands
5	—	Can lace shoes and is beginning to use tools Some are reading quite well and most can write their own names; questions about meaning of words
6	—	Reading Balls are bounced and (sometimes) caught

N.B. The information in this table is only approximate and does *NOT* fit every child. (For fuller details see Gesell, A., and Ilg, F. L. (1946) *The Child from Five to Ten*, London.)

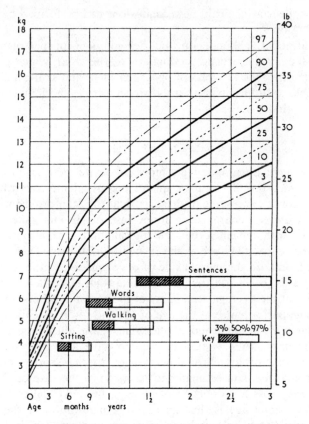

FIG. 41. Developmental landmarks and growth in weight (centile standards) during the first 3 years of life (English girls). (From *British Medical Journal* (1971) *2*, 125, by permission of the editor, Professor S. D. M. Court, and Dr. G. A. Neligan.)

The maturation of psychological awareness in the infant and child involves a progression from completely self-centred absorption to the recognition of the existence of others and finally to the development of an adult appreciation of the individual's place in society. During this period the child learns to cope with problems, and in this field boys are more aggressive than girls, who tend to ask for help more quickly, or to collapse into tears. On the other hand boys have a shorter attention span. It is important in relation to learning to remember that, although no new nerve cells appear after birth, new connexions can be formed between existing ones.

Additional evidence on the development of personality is afforded by the artistic productions of the child, which give an indication of the progress of his or her ideas and skills; it is often possible to differentiate sharply between male and female efforts of this kind.

MENTAL AGE

There have been many attempts to devise performance tests which measure the elusive quantity conveniently described as 'intelligence', and some of these, such as the Stanford–Binet Test, have achieved general recognition. Tests of this kind take cognizance of arithmetical, verbal, and logical ability, and also make use of other capacities such as the recognition of form. They roughly correspond, on the mental side of development, to the radiological atlases [p. 99] on the physical side; that is, certain test items represent the average attainment at the age of 9, others represent average performance at the age of 10, and so on. A child passing all the items in the group labelled nine years, but failing all those in the next group, would be said to have a mental age of 9, irrespective of his chronological age. Special scoring methods allow intermediate estimates to be made in the case of children who may pass a proportion of items in the next higher group.

The mental age is thus an index of maturation of the mind, and, like the radiological age, increases at a rate which depends on many intrinsic and environmental factors. Thus, unless the child has the opportunity to sharpen his wits on problems and to acquire experience, his mental age may lag behind, and, conversely, good teaching and interesting experiences may encourage his mental capacity to increase more rapidly. The fallacy that mental age proceeds at a steady rate independent of surroundings and ultimately attains a maximum level controlled only by intrinsic factors was one of the bases on which the 'eleven-plus' sorting test employed in British schools was used to direct children to various types of secondary schooling, and the virtual abandonment of this test indicates a recognition of the difficulties of drawing hard and fast distinctions by this kind of procedure.

So little is as yet known about mental processes that it is very difficult to say when mental development reaches maturity, but this event is thought to occur somewhere between 15 and 25 years of age. This is not to say that intellectual performance does not improve after this time, through the acquisition of knowledge and experience; the capacity of the mind to deal with intellectual material does not increase, but the 'know-how' due to experience enables it to function more smoothly.

It seems to be established that mental capacity, *as measured by tests of the kind described*, actually declines from the age of 25 or so onwards, and it certainly deteriorates markedly in old age.

A convenient way of classifying intelligence test performance is by the use of the concept of the 'Intelligence Quotient' (I.Q.)—a term which is now so familiar as to have entered everyday speech. The intelligence quotient is defined as the mental age divided by the chronological age and expressed as a percentage. Thus, a child with a mental age of 12 years and a chronological age of 10 would have an I.Q. of 120. Here again, as in the consideration of body size [p. 44], it is difficult to satisfy the natural desire of administrators to separate off 'normal' from 'abnormal'. A child with an I.Q. of exactly 100 obviously falls exactly in the middle of the 'standard' group from which the test was constructed, but what is to be made of a child with an I.Q. of 80? Is he to be described as simply of lower than average intelligence, or is he abnormal? Again, it is difficult always to be sure that like is being compared with like; I.Q. tests formulated in one country are not necessarily applicable to another, or even to a different set of children in the same country. The I.Q. is only a relative measure, not an absolute one, and it relates the performance of the child to that of a designated group of other children.

It may, however, be said that in Britain most university students fall into the group having intelligence quotients of 120 or over on the revised Stanford–Binet scale, while those with intelligence quotients of 60 or less are usually not capable of coping with ordinary education.

Another method of estimating mental development is simply to use as the standard the capacity of the child to read. A recent British survey of all children born in a given week in 1958 showed that by the time they had reached the age of 7 years several factors affecting the acquisition of skill in reading could be identified. For example, a child from socio-economic class V (unskilled manual workers) is fifteen times more likely to be a non-reader at this age than a child from socio-economic class I (higher professional workers). Again, those whose families live under poor or overcrowded housing conditions are behind those with no such handicap, and a child from a family of five or more children is about a year behind a child from a family of one or two, as well as being smaller [p. 155].

PHYSIOLOGICAL AGE

A whole series of interesting physiological and biochemical changes occurs during growth. Some of these changes are directly related to the alteration in the size of the child: for example, the resting heart rate slows progressively from about 100 beats per minute at the age of 2 years to the adult value of about 65–70 beats per minute. This falls in with the general biological rule that the heart rate is inversely related to body size. Similarly, the respiratory rate in a new-born infant can be as high as 40–45 per minute, compared with the adult rate of 12–16.

Some physiological functions 'mature' much earlier than others. The filtration rate of the glomeruli of the kidney reaches more or less its adult value by the age of about 2 or 3 years, but the blood pressure continues to rise, not only throughout the growth period, but also during the whole of adult life (the systolic pressure is about 80–85 mm of mercury at the age of 5 years, rising to the adult value of about 120 mm by the age of 18). The basal metabolic rate is highest in the new-born, and falls rapidly from 6 years to 20 years, with a more gradual fall throughout the rest of life. There is some doubt as to whether there is, or is not, a transient rise in the rate (or at least a failure to fall as expected) at the time of the adolescent spurt.

Many physiological and biochemical changes during growth show a sex difference in timing, for they are more closely related to other indices of maturation than to chronological age. Thus girls show a spurt in systolic blood pressure which occurs earlier than the corresponding spurt in the male, and the resting mouth temperature, which falls by $0.5-1.0\,°C$ from infancy to maturity, reaches its adult value earlier in girls. The red cell count and the blood volume of boys diverge away from the figures for girls at the time of the adolescent spurt [p. 88].

It has been suggested that some biochemical measurements might serve as an indication of physiological maturity, but there are considerable practical difficulties. Not only are there complications in measurement (for example, how does one define the 'resting state' in an infant 1 year old?), but there is so far a lack of the longitudinal studies which alone could provide a firm basis for such an assessment. It is known, for instance, that the affinity of haemoglobin for oxygen is greater in the new-born than in the adult, and it is said that there is a fall in this affinity between the ages of 2 to 15 years to values less than those of the adult, but as yet there is no firm evidence as to when the adolescent shift from 'low' to 'normal' values takes place. Perhaps a more promising index of maturity is the ratio of creatine to creatinine in the urine; this ratio is thought to fall progressively with age after about the age of $14\frac{1}{2}$ years, probably under hormonal influences. It has been shown that girls maturing early have a lower ratio than those of the same chronological age maturing late, and a measurement of this ratio might be made to afford information regarding maturity if considered along with skeletal and other data obtained at the same time.

The plasma inorganic phosphate shows a steady fall from the high levels of childhood to reach adult figures by the ages of 15 in girls and 17 in boys; the plasma calcium does not change during this period. The plasma alkaline phosphatase rises significantly, in parallel with the growth velocity, between the ages of 8 to 12 in girls and 10 to 14 in boys; thereafter it falls rapidly to adult levels.

LATE AND EARLY DEVELOPMENT

Six types of skeletal development have been recognized. There are first the average children, about whom little need be said. In the second group are the early maturing children who are tall in childhood only because they have matured faster than average: they will not be particularly tall adults. The third group comprises those children who are not only early maturers, but who are genetically tall also. These children are taller than average from early childhood, and will be tall adults. The fourth group is the opposite of the second, and contains children who are small because they mature late, but who will eventually be of average stature; the fifth group is the opposite of the third, and is made up of those who are both late in developing and genetically short in stature. Finally, there is an indefinite group who start puberty either much earlier or much later than usual. Since the 'growth life' of children in this category is longer or shorter than average [p. 27], they may well turn out to be much taller or much shorter than would have been predicted.

It is probable that terminal stature is related more to the mean growth rate during the whole growing period than to the time of onset of the adolescent spurt or to the time at which closure of the epiphyses takes place.

The evidence suggests that mental and psychological development are much more closely linked with radiological and dental age than with chronological age, and that an individual classified as 'advanced' or 'retarded' physically will also be so classified mentally. Those who are quick to mature physically score better in intelligence tests than those of the same age who have not yet reached the same stage of physical development. The difference ceases to be detectable after full maturation of both groups.

Both the late and the early developers have their problems. The child who develops late may lag not only in height and weight, but also in motor skills and in the development of intelligence. As a result he does less well in school and in athletics, and may be 'kept back' in a lower class than that appropriate to his chronological age. As a consequence of this he is likely to feel rejected, and he tends to be rebellious and aggressive, though his lack of muscular development renders him vulnerable in a fight. At the same time he is anxious about his failure to develop, and has a need for psychological support.

The child who develops early is in just as bad a case. To be physically bigger than one's colleagues is not necessarily an advantage when it comes to fitting in with the educational apparatus provided [p. 45], and the attainment of an adult body and sexual maturity before the attainment of economic independence produces a situation which is aggravated by the increasing amounts of secondary and tertiary education demanded by modern civilization. In this respect, as in so many other features of modern life, it is more comfortable to conform.

6

Changes in Shape and Posture

*'By reason of the frailty of our nature
we cannot always stand upright'*
Book of Common Prayer

THE ALLOMETRY EQUATION

Just as the whole body grows at a varying rate from birth to maturity, so its component parts have their own rates of growth, which also change with age, not necessarily synchronously. This leads to changes in bodily proportions.

Occasionally the relationship between the size of one part and the size of another, or between the size of one part and the whole body, can be described by a straight regression line; this argues that the relative growth rates of the two parts remain constant, and also that the difference between the two rates is established early in development.

But relations of this type may not always fit a simple linear regression, and differential growth is sometimes described by the so called allometry equation:

$$y = bx^k$$

where y is the size of one part and x that of another, and b and k are constants. The form of the allometry equation implies that the part of the body studied grows at a rate which increases or decreases proportionately with time (unless k is unity). There is no good evidence for this, and if the equation is used to study such matters as the relation between sitting and standing heights, arm length and trunk length, and so on, it is found that after a certain age new values for b and k become necessary for most paired comparisons. In other words, the relationship between one part and another of the growing body is not a steady or consistent one, but changes with age. A good example is to be found in the fact that sitting height increases at adolescence much more rapidly than stature as a whole, whereas previous to adolescence it more or less keeps pace with stature.

D'Arcy Thompson developed the theory of transformations to deal with the changing shape of the body produced by differential growth of its components. The outline of the animal or part being studied is projected on to a system of Cartesian co-ordinates, which can then be deformed mathematically, perhaps by tilting the axes, or by locally altering the scale in some parts of the figure but not in others.

SHAPE OF THE INFANT

It has already been said [p. 72] that the head of the new-born baby is relatively very large, and this is reflected in the finding that in the infant the ratio of vertical head height to total height is 1:4, whereas in the adult it is $1:7\frac{1}{2}$. This difference was not appreciated by many well-known early European painters, and it is common to find representations of the Virgin and Child in which the baby has the chubby outline typical of his age but the proportions typical of an adult.

Again, the lower limbs are much less well developed at birth than are the upper limbs: the ratio of lower limb length to total height is 1:3 in the new-born, but 1:2 in the adult. This means that as growth proceeds there is a shift in the centre of gravity downwards, from about the level of the 12th thoracic vertebra in the infant to about the level of the 5th lumbar vertebra in the adult. Since the child begins to walk before this shift has progressed very far, he is handicapped by being relatively top-heavy. The poor development of the lower limbs at birth is also shown by the fact that at this time they constitute only about 15 per cent of the total body weight, as compared with 30 per cent in the adult. At birth, the length of the foot is virtually the same as the length of the tibia, but in the adult it is only two-thirds as long. The span of the outstretched arms is less than total height at birth, but rapidly overtakes it [FIG. 42].

There are racial variations in these bodily proportions. For example, Negroes have relatively longer limbs, and their

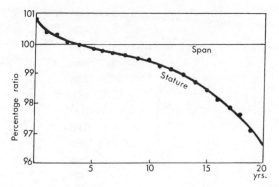

FIG. 42. Variation with age in the ratio of stature to the span of the outstretched arms. Note that the ratio is still decreasing at the age of 20. (From Thomspon, D'Arcy W. (1942) *Growth and Form* 2nd edn, London, Cambridge University Press, by kind permission of the publishers.)

forearms and legs are longer relative to their arms and thighs than they are in the white races.

Because of the differential growth of the lower part of the body, the umbilicus, which at birth is 1–2 cm below the mid-point of the body, has reached the mid-point by the age of 1 year; in the adult it lies higher still, about 60 per cent of the way up a line joining the heels to the top of the head.

The circumference of the skull at birth averages 33–36 cm, and this is about the same as the circumference of the chest: the circumference of the abdomen is greater than either until about the age of 2 years, because of the presence of the large liver [p. 91] and because the relatively small pelvis cannot contain several organs which are to be found in the adult pelvis; the best example is the urinary bladder [FIG. 36]. The rapid development of the pelvis in early childhood allows the bladder and intestines to sink down into it, and this results in a flattening of the abdominal wall. The thorax, which, with the shoulder girdle, has been 'pushed up' towards the neck by the overcrowding in the abdomen, is allowed to descend, and there is thus an apparent lengthening of the neck, which is usually not visible in a well nourished baby, the chin and lower jaw being in contact with the chest and shoulders.

The shape of the infant determines some of the hazards of infancy. For example, the shortness of the neck makes it difficult and dangerous to perform a tracheostomy, for the left

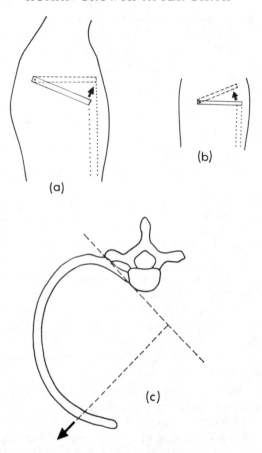

FIG. 43. Mechanics of respiration

(*a*) Diagram showing movement of adult rib. As the rib is raised, so the horizontal distance between its ends is increased, and thus the cavity of the chest enlarges until its size becomes maximal when the rib is horizontal.

(*b*) In a baby's chest the rib is already horizontal. Further elevation of its anterior end decreases rather than increases the anteroposterior diameter of the chest.

(*c*) Diagram to show that the motion of the anterior end of the rib is outwards as well as forwards. The rib articulates with the vertebra by two joints, one on the transverse process of the vertebra and one on the body, and rotates around a line drawn through these two joints. Its anterior end moves outwards and forwards in a direction at right angles to this line as the rib is raised to the horizontal, so enlarging the transverse as well as the anteroposterior diameter of the chest. Again, elevation of the horizontal rib in the baby cannot be effective in this way.

brachiocephalic vein may be pushed up above the suprasternal notch. The cross-section of the barrel-shaped infant chest is virtually circular, and remains so for the first two years; the ribs lie more or less horizontally. Now thoracic respiration depends on being able to raise the ribs to a more horizontal position from a more vertical one, for their anatomy ensures that this manoeuvre will increase both the antero-posterior and the lateral diameters of the chest [Fig. 43]. But if the ribs are already horizontal, it becomes mathematically impossible to increase the diameter of the chest by moving the ribs. It follows that the new-born baby has to depend almost completely on diaphragmatic breathing, and if an abdominal wound—for example a surgical operation—causes pain on movement of the abdominal wall, inadequate ventilation of the lungs may follow and pneumonia is a risk.

Because the rib cage and sternum are relatively much higher than in the adult, the abdominal cavity of the infant is less well protected by bone. Another anatomical danger lies in the fact that the external auditory meatus is poorly developed at birth [p. 79]. The ear drum is thus very close to the surface, and so is the facial nerve. Injury to the nerve may therefore be caused by obstetrical forceps, and damage to the drum by unwary clinical examination.

The size of the infant in itself constitutes a danger. Like all small animals, a baby has a relatively large surface area [p. 42], and the risk of losing heat is considerable. This is probably the reason for the plentiful supply of fat [p. 47], which acts as an insulating agent between the internal organs and the outside world. Conversely, because of the immaturity and inadequacy of the sweat glands [p. 50] the infant cannot adjust its temperature efficiently by sweating if it becomes overheated.

CHANGES IN SHAPE WITH GROWTH

Figure 44 and Table 6 show the alterations which occur during growth in the proportions of the various parts of the body. At all ages the dimensions of the head are in advance (i.e. nearer maturity) of those of the trunk, and the trunk is in advance of the limbs, and at all ages the more peripheral parts of the limbs are in

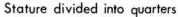

Stature divided into quarters

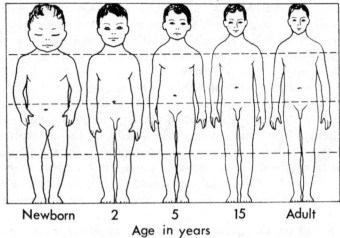

| Newborn | 2 | 5 | 15 | Adult |

Age in years

FIG. 44. Changes in bodily proportions with age. All figures are divided into quarters by the horizontal lines. Notice the great changes in the relative size of the head and the lower limbs, and how the mid-point of the body descends.

advance of the more central parts. Thus, the foot is nearer adult status than the calf, and the calf is more advanced than the thigh.

In the adolescent spurt, the feet and hands speed up first, then the calf and forearm, followed by the hips and chest, and then the shoulders. Last of all to accelerate are the length of the trunk and the depth of the chest; there is about a year between the peaks for lower limb length and trunk length. There is thus a transient stage during which the hands and feet are large and ungainly

TABLE 6

RELATIVE SIZES OF PARTS OF THE BODY

AGE	HEAD AND NECK	TRUNK	UPPER LIMBS	LOWER LIMBS
Birth	30	45	10	15
2 years	20	50	10	20
6 years	15	50	10	25
Adult	10	50	10	30

The figures are approximate percentages of total body volume.

relative to the rest of the body, and this is sometimes a source of anatomical embarrassment to the adolescent. The spurt in trunk length is relatively greater than the spurt in the lower limbs, so that more of the increase in height at adolescence derives from the trunk than from the lower limbs: the foot stops growing early, before almost all other parts of the skeleton.

The bones of the face grow faster than those of the cranial vault [p. 75], and it has been said that at adolescence the face 'emerges from under the skull' [FIG. 45]. In infancy the main growth is in the jaws, which have to accommodate the teeth, but in adolescence the profile becomes straighter and the nose more projecting. All these changes are more noticeable in boys.

At adolescence the shape of the face is also altered by changes in the hair. In the region of the frontal hairline the terminal hairs are replaced by vellus in about 80 per cent of females and in virtually every male, so causing a frontal recession of hair which may be progressive for several years.

In the later stages of the adolescent spurt there is a laterality of growth rather than a linearity, because the limb growth, which started first, exhausts itself first. The child who 'shoots up like a

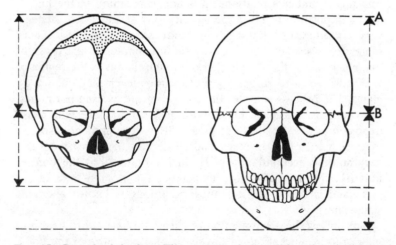

FIG. 45. Growth of the face. The newborn skull and the adult skull have been drawn so that the height of the cranial vault (the distance between the planes A and B) is the same. Notice the great relative increase in the facial skeleton in the adult. [See also Fig. 29.]

beanpole' proceeds to fill himself out in the later stages of his adolescent spurt [p. 34], and the clavicle, which thrusts the shoulder out from the trunk, is one of the last bones in the body to stop growing. Slow-maturing children tend to be long-legged and narrow hipped at maturity: fast maturing children become broader and more stocky.

During the adolescent spurt the sex differences in the skeleton become apparent; in the male the shoulders grow more than the pelvis, and in the female the reverse is true. The differential growth of the female pelvis, which causes it to become wider, shallower, and roomier than the male pelvis, is obviously related to the needs of childbearing, and the greater growth of the male shoulders is presumably related to the greater use of his more powerful muscles. But it is quite unknown how these changes are produced, though naturally the exciting agent is supposed to be the gonadal hormones [p. 147].

Other differences between the male and female skeleton are present at birth, and are merely exaggerated at adolescence. For example, throughout the whole period of growth the male forearm is longer, relative to height, than the female forearm. Again, the female index finger is often as long or longer than the ring finger, but this relationship is not so common in the male.

The deposition of fat in the female body at adolescence effects very considerable alterations in body shape [FIG. 19]. There seems to be no good reason why the fat should be laid down at one place rather than at another, although the fat in the region of the breast clearly serves to protect the developing mammary gland. The process is under the control of the ovarian hormones, and administration of these hormones to a male will cause deposition of fat in the breast as well as development of mammary tissue. In the Bushmen and Hottentots of Africa there is a very marked accumulation of fat in the buttocks at the time of puberty; this is much more evident in the women than in the men, and is known as steatopygia. Nothing is known about its significance.

Not only is the accumulation of excess fat harmful to health, it may produce alterations in the shape of other parts of the body. Thus, the increased weight may lead to flat feet or knock knees, or to an increase in the lumbar curvature of the vertebral column [p. 133], a condition known as lordosis.

CLASSIFICATION OF PHYSIQUE

It is well known that people have different shapes as well as different sizes. It is possible to give mathematical expression to this fact by using the complicated statistical technique of factor analysis, based on measurements made of various parts of the body. Such measurements may include total height, limb length, arm circumference, etc., and correlation coefficients can be derived from them. By factor analysis it is possible to account for much of the variability of the original measurements in terms of a smaller number of independent and more generalized factors. Such methods are as yet in their infancy, and, although they have a considerable potential value in research, they are not adequately descriptive of physique for the non-specialist.

Many attempts have been made to derive a meaningful classification of the variety of shapes presented by the human physique. The reason for this is that it appears that different physiques are associated with different diseases and with different temperaments. The matter is therefore far from being a purely academic problem.

One system at present in use is to give an 'androgyny rating'. This is a general assessment of the shape of the adult body in terms of sexual differences. For example, the male has a low and indefinite waist, whereas the female has a well marked one. The space between the legs in the male is relatively wide, and in the female is smaller. The distribution of body hair differs in the two sexes, and so does the amount of fat. These and other observations can be used to form a general rating system which gives some information regarding the shape of a given individual.

However useful this may be in a general way, it is not detailed enough to provide us with information from which clinical and physiological conclusions may be drawn. The system which is in most general use for this purpose is that devised by Sheldon and his collaborators, and is known as 'somatotyping'.

Sheldon recognized three different components of physique. The first is 'endomorphy'. An individual high in this component—an 'endomorph'—is characteristically round. He has a round head, his abdomen is larger than his thorax, and his arms and thighs contain much fat, though his wrists and ankles are relatively thin. He has large viscera, and a good deal of subcutaneous fat. His body measures more from front to back than from side to side.

A 'mesomorph', on the other hand, has strong shoulders and chest, a transversely lying heart, and heavily muscled arms and legs: his forearms and calves are relatively stronger than his arms and thighs. He has little subcutaneous fat, and is thicker from side to side than from front to back.

An 'ectomorph' is thin and narrow, with little muscle, little subcutaneous fat, and thin arms and legs. His heart lies vertically, and both his skin surface and his nervous system are relatively large. His adolescent spurt takes place about a year later than that of a typical mesomorph.

Sheldon found it possible to recognize the relative contributions to each of these three components of physique in any given individual, and so to record his 'somatotype'. By using a special photographic technique, he developed a subjective scoring system in which each component was allotted a mark varying from 1 to 7 arbitrary units. Thus, a person in whom ectomorphy predominates to an extreme degree might be recorded as 1-1-7, whereas an extreme endomorph would be written 7-1-1. These are naturally unusual somatotypes, and most people are classified round about 4-3-3 or 3-4-4 [FIGS. 46 and 47]. It is remarkable that with such apparently vague and subjective criteria, different observers, if properly experienced, vary by only about half a mark in relation to the analysis of each individual.

Difficulties are introduced into the statistical handling of somatotype data by the fact that the three components are not independent of each other, so that certain theoretically possible somatotypes are not found in practice. Thus there are no people who can be written 7-7-1 or 7-3-6, to take only two of many such examples.

The normal age for taking the somatotype is 20 to 25 years, and it is believed to remain constant throughout adult life. With the passage of years, an individual may put on or lose fat, increase the size of his muscles by training, or allow them to atrophy from disuse. Nevertheless it is claimed that a trained somatotyper can penetrate the various changes a person might make in this way to his basic external form.

Most of the work on somatotyping has been done with males, but women also show a similar range of physique. The distribu-

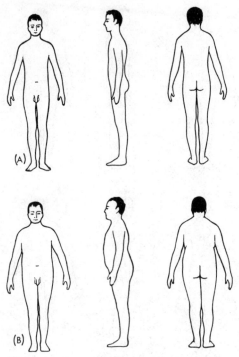

FIG. 46. Somatotypes

(a) An average individual, showing even balance between the three components of the somatotype 4–3½–4.)

(b) Predominant endomorphy, with minimal admixture of the other components. (Somatotype 7–1–1½.)

(Based on photographs in Sheldon, W. H., and Stevens, S. S. (1942) *The Varieties of Temperament*, New York, Harper and Row, by kind permission of the authors.)

tion of somatotypes is, however, somewhat different, for mesomorphic build is relatively much rarer in females than in males. The difference may be in part genetic, and in part because the muscular development dependent on testosterone [p. 147] does not usually take place.

Certain facts have accumulated regarding the relationship of body form measured in this way to physical, psychological, and physiological differences. For example, endomorphs more readily put on fat in middle age, and are more prone to develop a late onset diabetes than an early onset diabetes Ectomorphs are said

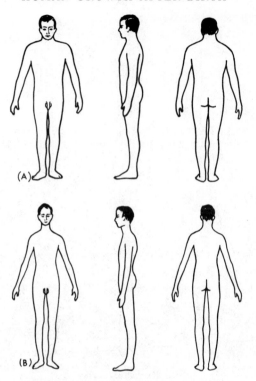

FIG. 47. Somatotypes

(a) Predominant mesomorphy, with minimal admixture of the other components. (Somatotype 1–7–1½.)

(b) Predominant ectomorphy, with minimal admixture of the other components. (Somatotype 1½–1½–7.)

(Based on photographs in Sheldon, W. H., and Stevens, S. S. (1942) *The Varieties of Temperament,* New York, Harper and Row, by kind permission of the authors.)

to have short reaction times to external stimuli, and are liable to tuberculosis. Mesomorphs have a high blood volume, and commonly suffer from coronary disease. The reasons behind such empirical observations indicate a fascinating line of research.

Studies in Africa have shown that the warmer the environment the lower is the body weight relative to height and the longer are the limbs relative to the trunk. Under the warmest conditions physique tends to become linear, as in the Dinkas [p. 43] who have limbs which are narrow and elongated (particularly the forearms and legs), a short trunk, and little subcutaneous fat. Whether this physique results from genetic, nutritional, or physiological influences [p. 157] is not yet established, but people with a linear build have a greater surface area than endomorphs or mesomorphs of similar body weight. It is therefore at least possible that the linearity of the Dinkas represents to some extent an evolutionary adaptation.

Sheldon and Stevens analysed personality in exactly the same way as physique, by describing three components of temperament; each of these was defined by twenty traits, which are shown in TABLE 7. Each trait was scored on a scale from 1 to 7, and the trait score was then averaged to get the rating for each component. When the personality ratings of a series of subjects were compared with their somatotypes, significant correlations (about 0·8) were obtained between 'viscerotonia' and endomorphy, 'somatotonia' and mesomorphy, and 'cerebrotonia' and ectomorphy.

Other studies have dealt with the effect of body type on choice of career. It is not unexpected to find that mesomorphs choose careers consistent with their athletic build—physical training instructors, army officers, and so on. But it has also been found that workers on the research side of a factory at all levels were more ectomorphic than those who were concerned with production in the same factory. In another investigation, students of engineering, medicine, and dentistry were found to be more mesomorphic than students of physics and chemistry, who tended to be more ectomorphic. Yet again, it appears that juvenile delinquents tend to be mesomorphic.

Others still have related different kinds of psychiatric disturbance to body build. Anxiety states seem to be more common in ectomorphs, though hysteria and depression are commoner in those with high scores in endomorphy and mesomorphy. Manic-depressive insanity tends to occur in people high in endomorphy and mesomorphy, but ectomorphs suffer from schizophrenia. There seems to be a pretty fair share-out of trouble all round.

It would clearly be important if we could predict the adult somatotype by means of information derived from measurements of the child. But at present children cannot be somatotyped with any accuracy, since there is an inadequate amount of information regarding transformations in body shape with age during the

TABLE 7

PERSONALITY TRAITS

VISCEROTONIA	SOMATOTONIA	CEREBROTONIA
*1. Relaxation in Posture and Movement.	*1. Assertiveness of Posture and Movement.	*1. Restraint in Posture and Movement Tightness.
*2. Love of Physical Comfort	*2. Love of Physical Adventure.	2. Physiological Over-response.
*3. Slow Reaction.	*3. The Energetic Characteristic.	*3. Overly Fast Reactions
4. Love of Eating.	*4. Need and Enjoyment of Exercise.	*4. Love of Privacy.
5. Socialization of Eating.	5. Love of Dominating, Lust for Power.	*5. Mental Overintensity, Hyperattentionality. Apprehensiveness.
6. Pleasure in Digestion.	*6. Love of Risk and Chance.	*6. Secretiveness of Feeling, Emotional Restraint.
*7. Love of Polite Ceremony.	*7. Bold Directness of Manner.	*7. Self-Conscious Motility of the Eyes and Face.
*8. Sociophilia.	*8. Physical Courage for Combat.	*8. Sociophobia.
9. Indiscriminate Amiability.	*9. Competitive Aggressiveness.	*9. Inhibited Social Address.
10. Greed for Affection and Approval.	10. Psychological Callousness.	10. Resistance to Habit and Poor Routinizing.
11. Orientation to People.	11. Claustrophobia.	11. Agoraphobia.
*12. Evenness of Emotional Flow.	12. Ruthlessness, Freedom from Squeamishness.	12. Unpredictability of Attitude.
*13. Tolerance.	*13. The Unrestrained Voice.	*13. Vocal Restraint, and General Restraint of Noise.
*14. Complacency.	14. Spartan Indifference to Pain.	14. Hypersensitivity to Pain.
15. Deep Sleep.	15. General Noisiness.	15. Poor Sleep Habits, Chronic Fatigue.
*16. The Untempered Characteristic.	*16. Overmaturity of Appearance.	*16. Youthful Intentness of Manner and Appearance.
*17. Smooth, Easy Communication of Feeling, Extraversion of Viscerotonia.	17. Horizontal Mental Cleavage, Extraversion of Somatotonia.	17. Vertical Mental Cleavage, Introversion.
18. Relaxation and Sociophilia under	18. Assertiveness and Aggression under	18. Resistance to Alcohol, and to

TABLE 7 (Continued)

VISCEROTONIA	SOMATOTONIA	CEREBROTONIA
Alcohol.	Alcohol.	Other Depressant Drugs.
19. Need of People When Troubled.	19. Need of Action When Troubled.	19. Need of Solitude When Troubled.
20. Orientation Toward Childhood and Family Relationships.	20. Orientation Toward Goals and Activities of Youth.	20. Orientation Toward the Later Periods of Life.

NOTE. The thirty traits asterisked constitute collectively the short form of the scale. (The table is reproduced from Sheldon, W. H., and Stevens, S. S. (1942) *The Varieties of Temperament*, New York, Harper and Row, by kind permission of the authors.)

period of growth. It is thought possible that in the future satisfactory predictions may be made when the child is about the age of 5 years or less, although difficulties are necessarily introduced by changes in shape occurring at adolescence.

CHANGES IN POSTURE WITH GROWTH

The vertebral column of the new-born infant has two primary curvatures which depend largely on the shape of the component bones [FIG. 48; PLATE 6]. Both are concave forwards, one being

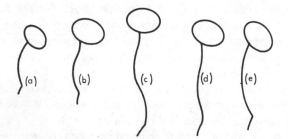

FIG. 48. Changes in the spinal curvatures with growth.

(*a*) Infant. Two primary curvatures.
(*b*) Six months. The secondary cervical curvature has appeared.
(*c*) Adult. Two primary and two secondary curvatures.
(*d*) Old age. The two secondary curvatures, dependent on the discs and the postural muscles, are becoming obliterated.
(*e*) Final stage, corresponding to condition in infant.

in the thoracic region, and the other being formed by the curve of the sacrum, which at this stage consists of separate sacral vertebrae. Later on the sacral curvature becomes permanently fixed by the fusing together of its vertebral components; the thoracic curve allows a certain limited amount of movement, but the intervertebral discs in this region are thin, and this restricts the scope of movement of one vertebra on another; movement is also limited by the oblique set of the spines of the vertebrae and the fact that their laminae overlap each other.

At about 3 months of age the baby begins to hold its head up, and in association with this a secondary curvature appears in the cervical region of the column. This curvature, unlike the two primary ones, is convex forwards, and remains mobile, for its radius depends on the tension of the muscles which stretch across its concavity. The mobility of the cervical curvature is largely due to the thick intervertebral discs, which allow a considerable 'play' between one vertebra and its neighbours.

On top of this secondary curvature the skull has to be held balanced. At birth the position on the skull of the facets which articulate with the atlas vertebra is similar in all anthropoids. In apes the portion of the base of the skull in front of the joint grows more than the portion behind, so that the joint shifts backwards and the centre of gravity of the skull and brain falls well forward. In man there is much less discrepancy between the growth of the skull in front of the joint and behind it. The result is that the joint comes to lie relatively much further forwards than in the apes, although the centre of gravity of the head is still in front of it. The skull therefore balances reasonably well on top of the vertebral column, but it still requires some muscular effort to keep the gaze horizontal.

When the baby begins to sit up, the lumbar curvature appears. This secondary curvature, like the cervical one, is mobile, depends largely on the discs rather than on the shape of the bones, is controlled by the great postural muscles of the vertebral column, and is convex forwards [FIG. 48]. Both the secondary curvatures may fail to develop at the expected time should there by any delay in the development of the postures of sitting and holding the head up.

At the creeping stage the baby is essentially a four-footed animal, albeit with exceptionally mobile forelimbs, and his centre of gravity is supported in a very stable manner by his quadrupedal posture. When this quadruped rears himself up on end in his first efforts to walk, he has an awkward and unstable stance. Since his centre of gravity is high [p. 118], he must stand with his legs widely apart in order to balance himself securely, and since it lies well forward, partly because of the large liver [p. 91], there is a compensatory exaggeration of the lumbar curvature in order to bring the upper part of the body into the vertical position; this usually ceases to be necessary about the age of 4 years. In the early stages of walking the baby is often bow-legged [Fig. 49]; this corrects itself gradually, and may be followed by a knock-kneed period. About the age of 3–3½ years 75 per cent of children have a distance of more than 2·5 cm between their medial malleoli, and in over 20 per cent the distance is 5 cm or more. The condition usually rectifies itself by the age of 7 years. As the baby stands up, so the weight of the body pressing down on the lumbosacral joint forces the upper part of the sacrum forwards and downwards; the posterior part of the sacrum and the coccyx are restrained by the sacrotuberous and sacrospinous ligaments from rotating upwards in consequence of this, with the result that the sacrum sinks more deeply in between the two hip

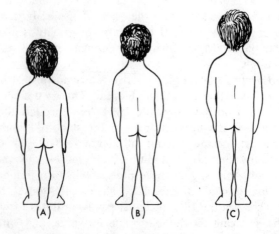

FIG. 49. Development of stance.

(a) 18 months old: bow legs. (b) 3 years old: knock knees.
(c) 6 years old: legs straight.

bones. As the legs straighten and the weight is transmitted through the pelvis to the heads of the femora, the pelvis becomes steadily remodelled; the force applied to the sacrum tends to lever outwards the lower part of the pelvis, thus broadening the region of the symphysis and increasing the subpubic angle. The acetabula become deeper and the hip joints more stable as the child becomes more active in his newly acquired two-legged freedom.

In the new-born the legs are flexed and the feet inverted; as walking begins the lower limbs straighten out and the feet evert, and at the same time the angle made by the neck of the femur with the shaft decreases gradually from about 160 degrees to the adult value of about 125 degrees. The shaft of the infant femur is straight, and the forward curve characteristic of the adult bone begins to appear as the child stands up.

In the adult the centre of gravity is approximately 55 per cent of the total height from the floor, being higher in men than in women. The line of gravity naturally alters constantly according to posture. When standing at rest it falls through the external auditory meatus, behind the hip, in front of the ankle, and half-way between the heel and the balls of the toes [Fig. 50]. It passes through the dens of the axis vertebra, the front of the body of the second thoracic vertebra, the middle of the body of the twelfth thoracic vertebra, and the back of the body of the fifth lumbar vertebra [Fig. 51]. If the centre of gravity is allowed to fall too far backwards or forwards, muscular effort is needed to prevent falling.

The most important single factor in faulty posture is the tilt of the pelvis in relation to the horizontal. This is defined as the angle with the horizontal made by a line through the sacral promontory and the upper border of the symphysis pubis [Fig. 51], which is normally about 60 degrees. The pelvic tilt is determined by the postural pull of the muscles of the back, abdomen, and thighs, and these pulls are in turn influenced by the way the individual habitually stands. A 'tense' habit of standing increases the tilt, so that the pelvis rotates forwards on the thighs, carrying the lumbar spine forwards, and with it the centre of gravity. In compensation, the upper part of the body is thrust backwards, so increasing the lumbar curvature. The neck is held stiffly, with the chin tucked in, in order to maintain the

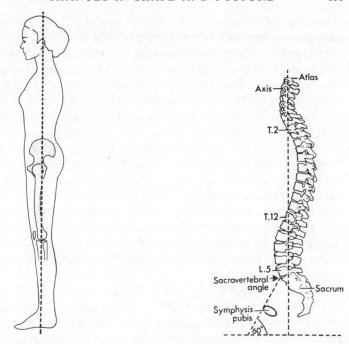

FIG. 50. Position of line of gravity in erect posture. The line passes through the external auditory meatus, the tip of the acromion process of the scapula, and the mid-point between the heel and the balls of the toes. It falls behind the hip joint, through the middle of the knee joint, and in front of the ankle joint.

FIG. 51. The line of gravity. The line passes through the dens of the axis vertebra, the front of the body of the second thoracic vertebra, the middle of the body of the twelfth thoracic vertebra, and the back of the body of the fifth lumbar vertebra. The angle made with the horizontal by a line passing through the sacral promontory and the upper border of the symphysis pubis is known as the pelvic tilt, and is usually about 60°.

horizontal gaze of the eyes. If this posture is held as a routine, the muscles which have pulled the pelvis out of position may shorten and their opponents lengthen, so that after a time the individual is unable to stand in any other way.

The converse of the 'tense' posture is the 'slack' posture. The pelvic tilt decreases, the centre of gravity passes backwards, and the head and thorax are thrust forwards to compensate. There is thus a mild increase in the thoracic curvature, and the neck is extended, the chin being poked forwards. The joints of the lower limb tend to flex, and become mechanically unstable.

Poor posture results from many causes, which are not always as easy to determine as might be thought. The damage may originate in any part of the body, for posture is an integral whole, and anything that tends to upset one part of the mechanism will throw the rest of it out of gear. A good example is the very common defect known as 'round shoulders'. By this is meant a failure of the muscles of the shoulder girdle to hold the scapulae back towards the spine, so that the whole shoulder girdle drifts round the side of the chest towards the front of the body. This in turn upsets the gravitational equilibrium of the upper part of the body, for arm weight in front of the gravity line must be compensated for by alterations in the curvatures of the spine, and this in turn causes a movement of the pelvis which has repercussions on the posture of the joints of the lower limb and on the distribution of weight in the foot.

Conversely, the wearing of high-heeled shoes, which tip the body weight forwards, can lead to disturbances of the posture of the whole body working upwards through the pelvis to the spine; a complaint of aching pain in the neck can sometimes be cured by wearing a lower heel.

Postural deformities can obviously be caused by injury or disease, but there is also a hereditary factor, and mental attitudes are also of importance in determining how the child 'holds himself'. Deformities commonly develop in adolescence, when the weight increase at the time of the adolescent spurt may not be accompanied for some time by a corresponding increase in the strength of the postural muscles, especially in girls. Those people who have six lumbar vertebrae instead of five have a greater part of their spinal column unprotected by the leverage of the rib cage, and tend to have trouble in this region. One of the most difficult conditions to explain is scoliosis, which is a twisting deformity affecting the whole column. It is often attributed to faulty working postures in poorly designed school seats and desks, but there are certainly other factors concerned.

DRAWBACKS OF THE ERECT POSTURE

The erect posture brings with it certain advantages. The hands are freed, allowing manipulation of objects and co-ordination between the hands and brain. This in turn is responsible for an increase in the size of the brain, and thus for our capacity to make and use tools. A second advantage is that the eyes are brought further from the ground and made more mobile, since the skull can now be poised more freely on the top of the vertebral column.

But the erect posture also brings with it many disabilities, some of which are apparent quite early in life, while others tend to appear later, when the pull of gravity and the effects of tissue degeneration have been operating for a long time.

In the first place, it is a precarious equilibrium, and consequently a great part of the large central nervous system has to be devoted to maintaining this equilibrium by a complex system of reflexes and controls. These are easily lost as a result of illness, or, as we shall see [p. 212], in old age. Secondly, there is an increased strain on the bones and joints of the lower limb, particularly on the foot, and this is shown by the incidence of flat feet, sprained ankles, etc. These are much in evidence following the adolescent spurt, when violent exercise may throw an intolerable strain on the ankle region, and when periods of prolonged standing at work may place too much gravitational load on the arches of the feet.

There is also a considerable strain on the cantilever structure of the vertebral column, which is not yet fully adapted by evolution to being tipped up on end [Fig. 52]. Aches and pains, degenerating joints, and damage to intervertebral discs are all common in the lower part of the column.

The pelvis is also not yet properly adapted, and remains essentially the pelvis of a four-footed animal. Nevertheless, it has to continue to allow the passage of the fetal skull, which in the course of evolution has grown bigger, while at the same time coping with a very large gravitational strain. This mechanical problem is aggravated by the postural burden which a heavy pregnant uterus [p. 36] lying in front of the line of gravity imposes on the vertebral column, and women who have borne several children may have troublesome pain in the lower part of the back resulting from the stress laid upon their lumbosacral and sacro-iliac joints.

In the upright position the forelimbs hang down at the side of the trunk, at right angles to their former 'neutral' position. The

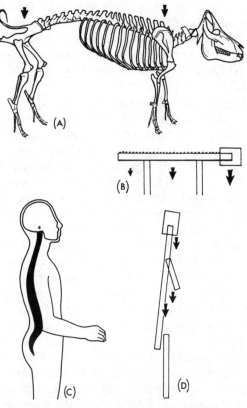

FIG. 52. Stresses in the vertebral column.

(a) Skeleton of a pig: the arrows indicate the points at which weight is transferred to the limbs.

(b) Compare with the plan of a two-armed cantilever. The solid line indicates compression forces taken by the bones, the dotted line indicates tension forces. The spines of the vertebrae in (a) are long and powerful in order to afford attachment to muscles and ligaments resisting tension forces.

(c) and (d) When this structure is stood on end, the stresses which it has to withstand are quite different.

resulting extensive anatomical alterations in the shoulder region have made some of the nerves and vessels entering the limb liable to compression and injury.

Breathing is hampered, since the whole weight of the chest wall must be raised against the pull of gravity instead of swinging at right angles to it as in the four-footed animal. The rib cage therefore tends to fall in old age, when the muscles are no longer equal to the heavy task of raising it in inspiration; this is responsible for respiratory difficulties due to deficient ventilation. At the same time the thoracic and abdominal viscera descend, and the anterior abdominal wall becomes weakened because of increased pressure on the lower portion of it; this in turn is a factor in the frequency of hernia. Similarly, there is an increased gravitational load on the pelvic floor, and this shows itself in the frequency of the condition of prolapse, in which the uterus and bladder descend through a pelvic diaphragm weakened or damaged by childbirth.

Finally, the elevation of the brain above the heart means that gravity interferes with its blood supply. The return of blood to the heart from the lower limbs is also impeded by the long vertical haul against gravity. For this reason the autonomic nervous system has had to develop a complicated system of vasomotor controls, and these are not always adequate to the occasion, as is seen in the condition of postural fainting, in which the blood supply to the brain is interrupted when the patient suddenly stands up. The common condition of varicose veins also testifies to the incomplete adaptation of the vascular system.

The erect posture is therefore not an unmixed blessing, and the gradual achievement of it during the first part of childhood carries with it liabilities as well as advantages. In old age many of these liabilities come home to roost, and posture undergoes regressive changes which will be dealt with in CHAPTER 10.

7

Factors Influencing Growth and Maturation

*'Boy (20), 5 ft. 7 ins., wishes
to increase height to 5 ft. 10
ins.'*

The Times, 6 Nov. 1968

GENETIC CONTROL

Both genetic and environmental factors influence growth, and the progress of any given child is the result of a complex interaction of many different factors. Nevertheless, the best way of growing tall and heavy is to have tall and heavy parents. Studies of twins have shown that body shape and size, deposition of fat, and patterns of growth, are all more nearly related to nature than to nurture. Heredity affects not only the end result of growth, but also the rate of progress towards it, and thus the radiological, dental, sexual, and neurological ages [CHAPTER 5] of identical twins tend also to be identical, whereas those of non-identical twins may differ considerably. There is also a strong genetic component in the determination of body shape [CHAPTER 6].

The control of body size is certainly a complicated affair involving many genes, yet a disturbance in a single gene or group of genes may produce a widespread and drastic effect, as in the condition of achondroplasia [p. 186], which is inherited as a simple dominant. On the other hand, the effects may be quite restricted and specific. The genetic control of dental maturation and eruption appears to be separate from that of skeletal maturation, and there is even evidence that the genes controlling the growth of different segments of the limbs are independent of each other.

It is now believed that dental development and the *sequence* of ossification are primarily genetically controlled; the *timing* of ossification is partly influenced by genetic factors and partly by environmental ones. Maturation as a whole is even more affected by environment, but genetic influences are still detectable.

Certain peoples are taller than others: the Dinkas are one of the tallest races, and the pygmies of the Congo one of the smallest. It is probable that genetic factors are largely responsible for such racial differences, though nutrition [p. 149] may also play an important part. It is now established that, at least in Central African pygmies, the secretion of somatotrophin by the pituitary [p. 144] is normal in quantity, and, so far as can be ascertained, in chemical composition. The presumed genetic effect which keeps them small must therefore act on the tissues upon which the hormonal control acts, rendering them in some way unresponsive; injection of somatotrophin into pygmies produces none of the usual biochemical changes associated with such injections in other races.

There is also interesting evidence of racial differences in skeletal maturation. Careful studies have shown that Negro children in West Africa, East Africa, and the United States are ahead of white children living in these areas at birth, and also for the first year or two after birth. The Negro babies grow faster at birth than the white babies, and they are also ahead in such important functional stages of development as sitting up and crawling [TABLE 5, p. 110]. However, the growth curves are different in the two races, so that by about the fourth year the initial difference in growth has been abolished, and by this time also the skeletal advancement, together with the advancement in performance tests, has disappeared. This might possibly be because by the third year the effects of inadequate nutrition have neutralized the racial or genetic advantage of the Negro children. But the permanent teeth also erupt about a year earlier in Negroes than in whites, although by this time the effects of nutrition would certainly be operative.

It is often supposed that the differences in average height between Eastern and European races are due to earlier closure of the epiphyses of the long bones in the former, but work in India and elsewhere has shown that there is insufficient evidence for this conclusion, and that these differences are probably due to differences in average growth per annum during the whole growing period, genetically determined, and modified by nutritional factors.

It would be interesting to have observations on the skeletal maturation and epiphyseal closure of pygmies, but the major difficulty in such an investigation is that of establishing the chronological ages of the subjects. The pygmies of New Guinea may possibly supply information in the near future.

Genetic factors probably play the leading part in the difference between male and female patterns of growth. As regards size, there is little to choose between boys and girls up to the age of 10 years, though boys are on the average slightly heavier at birth [p. 32]: it is the timing and intensity of the adolescent spurt which is responsible for the size difference between adult men and women, and the onset of the spurt is probably genetically controlled, although its intensity and duration probably depend on hormonal factors.

The marked and consistent advancement of girls over boys in respect of skeletal maturation [p. 101] has been attributed to a retarding action of the genes on the Y-chromosome of the male. Individuals with the chromosome pattern XXY suffer from what is called Klinefelter's syndrome; they are long-legged, show some enlargement of the breasts, and have grossly deficient spermatogenesis. Yet their skeletal retardation approximates to that of normal males, despite their possession of two X chromosomes, which, it might be imagined, would tend to produce a female pattern of maturation. Again, people with one instead of two X chromosomes (Turner's syndrome) develop as females with rudimentary or absent ovaries and lacking secondary sexual characteristics. Such people are usually plotted below the 3rd centile of height throughout growth, and do not grow to more than about 152 cm (5 feet) tall. The adolescent spurt does not occur, and the growth rate may actually decrease instead.

Individuals with the pattern XYY tend to be tall, and, it is thought, may have a tendency to violent or to criminal activities. Nothing much is yet known about the skeletal maturation of such people, though some children with this defect have now been identified.

NEURAL CONTROL

It has been suggested that there may be a 'growth centre' in the brain, possibly in the hypothalamus, and that this is responsible for keeping the child on his genetically determined growth curve wherever possible. If he deviates from this curve for any reason, such as malnutrition [p. 149] or illness [p. 158], a period of accelerated 'catch up' growth ensues. This phenomenon implies some sort of central control mechanism, and the reason for locating the hypothetical 'growth centre' in the hypothalamus is that the hypothalamus interacts with the anterior lobe of the pituitary gland in one of the basic hormonal controls of growth [p. 144].

There is also evidence that the peripheral nervous system may play some part in the control of growth. If somatic muscle is denervated it atrophies [p. 200], and, similarly, taste buds deprived of their innervation will degenerate. When a nerve supplying the hand or foot is cut, the growth of the nails within its territory is retarded in comparison with that of the nails supplied by other nerves, and returns to normal with the regeneration of the nerve. These effects and others like them are not sufficiently explained by the disuse following the injury, or by a diminution in blood flow consequent on this disuse. It is therefore suggested that the peripheral nerve fibres exert a nutritive or 'trophic' effect on the structures they supply, and that this is due to a chemical secreted by the nerve cells and liberated at the endings of the nerve fibres; the chemical in some way modifies the growth and repair pattern of the structures innervated. Acetylcholine has been incriminated as the neurotrophic agent, but sensory nerves, which produce comparatively little acetylcholine at their peripheral endings, exert a stimulating effect just as great as motor nerves, which produce much more. The present view is that the substance involved is not one of the normal neural transmitters, but much more evidence is required before its nature (or even its existence) can be regarded as settled.

Finally, it is interesting that regeneration of a limb in animals which are capable of this feat depends, in its early stages at any rate, on the presence of an intact and sufficiently dense sensory nerve supply to the stump [p. 170].

HORMONAL CONTROL

Probably all the endocrine glands influence growth. Fetal hormones can play no part in growth control before the end of the second month of fetal life, for until about this time the glands to elaborate them have not been formed. The maximum rate of growth in height occurs about the fourth month of fetal life [p. 21], and by this time the pituitary and thyroid glands are certainly functional. The gonads apparently play no part in fetal growth, and the activity of the parathyroids is uncertain. The role of the fetal suprarenal cortex is not yet fully established. However, it appears that the pars intermedia of the pituitary gland is responsible for breaking down adrenocorticotrophic hormone [p. 146] produced by the anterior lobe into a fraction which stimulates the fetal cortex to produce growth stimulating androgens [p. 146] rather than corticosteroids. Towards birth the pars intermedia becomes less active, the fetal cortex decreases in size, and the production of androgen falls while that of corticosteroids rises. It is therefore possible that this mechanism plays an important part in growth before birth.

The anterior lobe of the pituitary gland produces, among other hormones, a polypeptide which is called growth hormone, or somatotrophin. This can be detected in the body of the fetus as early as the end of the second month, very soon after the pituitary has been formed. Yet it does not seem to be essential for the growth of the fetus, for the bodies of anencephalic children, in whom the pituitary fails to develop, are of normal size. It was formerly thought that children deficient in somatotrophin could not be identified for several years after birth, since their growth could proceed apparently normally for some considerable time. But some at least of these children show a failure to grow which can be detected even in the first year of life, and it is now believed that somatotrophin is necessary for normal development right from the beginning of independent existence.

Somatotrophin maintains the normal rate of synthesis of protein in the body, and it also inhibits the synthesis of fat and the oxidation of carbohydrate. It is necessary for the proliferation of the cartilage cells of the epiphyseal plates [p. 62], and so exerts a great effect on height. In contrast, it has little apparent effect on the maturation of the skeleton [p. 198]. The daily output of somatotrophin secretion does not appear to vary with age, with the stage of puberty, or with the levels of gonadal secretion [p. 147] in the blood. The adolescent spurt does not depend on somatotrophin, and its influence on growth seems to become ineffective at the time the epiphyses close, though its metabolic effects probably continue to be felt throughout adult life, particularly as regards protein synthesis.

Somatotrophin operates, at least in part, through the intermediation of three, or possibly more, secondary substances called somatomedins. These are peptides which are formed in the liver and circulate in the blood plasma. They have several insulin-like effects, which include the stimulation of protein synthesis and the depression of protein breakdown, and indeed it has been said that in any list of growth-promoting peptides insulin itself must come first.

It is probable that there is a complicated interaction between somatotrophin and insulin. There is evidence suggesting that in some cases of diabetes an excess production of somatotrophin may be a factor in depressing insulin production by the islets of Langerhans, and there is an association between the condition of acromegaly [p. 188] and diabetes.

It is also likely that there is an antagonism between the production of somatotrophin and the production of steroids by the suprarenal glands. Patients on long-term corticosteroid therapy suffer an interference with growth, and during such treatment it may be impossible to stimulate growth by administering somatotrophin in doses far greater than those necessary to produce a response in children with deficient pituitary secretion of the hormone. Recovery of growth can occur when the dosage of corticosteroids is reduced. The amount of somatotrophin in the blood stream at a given time is subject to wide variation, and there appears to be a daily rhythm of secretion, the amount varying inversely with the amount of steroid secretion. Secretion of somatotrophin is subject to an extremely complex system of checks and counterchecks, feedbacks and controls. The hypothalamus produces a somatotrophin-releasing factor, which is probably secreted in response to a fall in the blood sugar, and a somatotrophin release inhibiting factor called somatostatin. The outpouring of somatotrophin is also influenced by such things as the intake of food and the amount of exercise. Like corticosteroids, somatotrophin is secreted intermittently. During sleep, and often during the day as well, bursts of secretion lasting up to two hours may occur, and between these bursts the hormone may be impossible to detect in the blood plasma. Secretion during sleep is associated with stage 4 (deep) sleep, and 20–40 per cent of the total 24 hours output occurs in the first 90 minutes of nocturnal sleep. Active secretion appears to take place during only about six hours of each day; the lower concentrations of hormone found at other times are probably the result of slow removal of hormone secreted earlier.

The anterior lobe of the pituitary gland also secretes, quite independently, a thyrotrophic hormone, which affects growth by stimulating the thyroid gland to secrete. The hormones produced by the thyroid gland are thyroxine and tri-iodothyronine, both of which stimulate general metabolism, and are particularly important in relation to growth and maturation of the bones, teeth, and brain. If there is a serious deficiency of iodine in the maternal diet, the fetal thyroid may be unable to produce normal amounts of secretion, and the brain fails to develop properly. The amount of thyroid secretion is thought to decrease somewhat from birth to puberty, at which time there is an interruption of the decline for the period of the adolescent spurt. If there is a deficiency of thyroid secretion in childhood, the growth of the whole body suffers, and the child becomes a mentally deficient dwarf [p. 191].

It seems to be agreed that the pituitary and thyroid glands play little direct part in the adolescent spurt, and this idea is supported by the abrupt change in the whole pattern of growth at this time—for example, the sudden appearance of the secondary sex characteristics, the closure of the epiphyses, and so on. These changes are attributed to two new factors, the first of which is that the cortex of the suprarenal glands begins to secrete substances called androgens.

The way in which this happens is not understood; the suprarenal cortex is under the control of adrenocorticotrophic hormone (ACTH) produced by the pituitary gland, and has since birth been routinely secreting corticosteroids at a more or less steady level. There is no change in the amount of secretion of ACTH during adolescence, and the other secretions of the suprarenal cortex continue steadily on. What, then, causes the cortex to produce an additional secretion? It has been suggested that some kind of inhibiting mechanism which prevents the androgens being liberated is removed at the time of puberty, but so far this is speculation. Whatever the mechanism, the androgens play a major part in determining the course of the adolescent spurt in both sexes.

The second factor is that just before the adolescent spurt the hypothalamus stimulates the pituitary to begin the production of gonadotrophins, which in turn stimulate the interstitial cells of the testicle or ovary to secrete. The gonads of both sexes secrete oestrogens in small quantities from the time of birth onwards. At puberty the oestrogen level rises sharply in girls and to a much more limited extent in boys; the sex difference is possibly due to an inhibitory hormone secreted by the seminiferous tubules of the testicle. Testosterone, produced by the testicle, is important in stimulating growth, as are the androgens, and the steadily rising output of testosterone is responsible for the greater growth of muscles and the greater number of red blood cells per unit volume in the male. The secretions of the ovary have apparently less effect on growth in general, and it follows that in a boy there are two lots of powerful growth-promoting chemicals circulating during the adolescent spurt, while most of the growth in a girl at this time depends on the androgens of the suprarenal cortex alone. The ovarian secretions are, however, powerful in a different direction, for they largely control the secondary sex changes in the female, including the alterations in the shape of the body [p. 124]. Just how the ovary influences differentially the osteoblasts and osteoclasts of the shoulder and pelvic regions has still to be explained, for there are no obvious anatomical differences in the two regions which could account for the effect. Nor is it clear how fat is deposited in certain areas of the body but not in others.

The level of oestrogens in the blood is thought to influence the timing of the closure of the epiphyses, and girls who are at risk of becoming excessively tall are sometimes given oestrogens to accelerate the termination of growth [p. 148]. In both sexes it is the androgens which stimulate the development of secondary sexual hair. In the female this is confined to the pubes and the axillae, where the skin is most sensitive to stimulation by androgens, but in the male it extends to adjoining areas of skin and to other regions such as the face and neck.

Finally, the parathyroid glands secrete parathormone, which withdraws calcium from the bones to maintain the concentration of calcium in the blood plasma at a constant level. In this it is opposed by another hormone, calcitonin, which is produced by the parafollicular cells of the thyroid gland and also to some extent by the parathyroids and the thymus. Calcitonin inhibits the drain on bone calcium, particularly from bones which are metabolically very active, as they are while they are growing; it is thought to reduce the numbers of osteoclasts and increase the numbers of osteoblasts.

Girls of 6 to 10 years of age mature skeletally much faster than boys, but in girls of 10 to 15 years the process decelerates relative to boys in the same age group. These stages of maturation are probably endocrine-controlled. It seems to be indicated that the replacement of cartilage by bone is controlled by the thyroid secretion until the time of puberty, but that after puberty the increasing influence of the gonadal hormones overtakes the action of the thyroid, and is responsible for the subsequent maturation patterns of girls and boys. Dental maturity, like skeletal maturity, probably depends on the thyroid gland until the spurt begins, and it is therefore not surprising that dental and skeletal maturity are usually well correlated with each other.

Should there be an excessive production of sex hormones in the growing child, sexual maturation may be considerably advanced. Growth in sexually precocious children usually ceases early, about the age of 10 years, so that they become smaller adults than those with a more normal timetable of development [p. 193].

Hormonal treatment is now used in an attempt to modify growth if excessive tallness or shortness can be confidently predicted from the bone age and other factors [p. 31]. To limit height androgens are given to boys and oestrogens to girls. Such treatment precipitates the appearance of the secondary sex characteristics, and hence should not start before the age of about 10. If the radiological age of a girl is over 13 or that of a boy is over 14, little success can be expected, but earlier than this some diminution, perhaps 4–5 cm, in the predicted height is possible. The object of treatment is to accelerate skeletal maturation so much that the time left for the child's height to increase is much reduced. However, androgens (though not oestrogens) have the drawback that they tend to cause an initial acceleration of growth which may partly nullify the total effect.

Because of the risk of thrombosis inherent in the administration of large doses of oestrogens it is probably not justifiable to use them except in very clear cut situations. An alternative is to recommend a surgical operation on the lower limbs; as much as 5–10 cm may be removed from height in this way without causing obvious deformity, because very tall children tend to have somewhat disproportionately long legs.

No surgical alternative exists for improving the height of small children, and hormonal treatment is also disappointing, unless the child has a deficiency of somatotrophin, insulin, or thyroid secretion. In such cases the results of specific replacement of the missing secretion can be remarkable [p. 189]. Neither somatotrophin nor thyroxin has any effect on retarded or stunted growth unless there is a pre-existing deficiency. Androgens in small doses have been given to increase height, but the risk is that exactly the reverse effect may occur if the dose is too large [see above]. It seems likely that administration of anabolic steroids [p. 36] never increases height to anything more than would eventually have been reached without them.

NUTRITION

A sufficiency of food is essential for normal growth, and in post-war Germany, under near famine conditions, children lagged ten to twenty months behind normal growth performance. Human malnutrition naturally involves not only a deficiency of calories, but a lack of specific foodstuffs, and it is impossible to separate the two factors. Several experiments on supplementation of inadequate diets have been carried out under conditions of widespread malnutrition, but most of our knowledge on the relation of nutrition to growth derives from animal work.

Some interesting experiments with rats and mice have shown that when fed on a diet deficient in calories but otherwise satisfactory they cease to grow, but can resume growth when an adequate supply of calories is restored. Rats have been kept immature for up to 1000 days on such a diet, and were still juvenile while their fellows on a normal diet were senile. On starting normal feeding, the rats rapidly grew to adult size, although those which had been kept on the deficient diet for more than three years could not always begin growing again. Subsequent work has shown that the diet which results in the maximal final size of experimental animals is one which, as judged by appetite and by growth rate, represents a moderate degree of under-feeding.

An adequate supply of calories is naturally essential for the normal growth of humans as well as rats, and the need varies with the phase of development. For example, in the first year of life a baby requires about twice as many calories per unit weight as a male adult doing moderately heavy work. Again, a schoolboy going through his adolescent spurt may require perhaps 3000 Calories a day—as much as a moderately heavy manual worker. He also needs about 50 per cent more calcium, nitrogen, and vitamin D than he did before the spurt began.

After a period of human or animal malnutrition has ended, it is found that growth accelerates in an attempt to 'make up' the loss in height and weight. Such an acceleration can be successful in restoring the predicted height if the period of malnutrition has been short, but sometimes complete compensation proves impossible. The phenomenon of growth acceleration in such circumstances raises interesting questions regarding the central control of growth [p. 143], and the capacity of the body to return to a genetically determined growth curve after having been pushed off it has been termed 'homoeorrhesis'.

Undernutrition at any given stage of development tends to accentuate the normal process of differential growth [p. 68]. For example, in dietary deprivation the growth of the teeth takes precedence over the growth of the bones, and the bones grow better than the soft tissues such as muscle and fat. Skeletal maturation is retarded less than skeletal growth, and in the brain myelination is less affected than total brain growth. Again, under poor dietary conditions the growth of the sexual organs at puberty is relatively less depressed than that of other tissues and organs. Restriction of food slows down production in the germinative layer of the skin, but also decreases the rate of loss from the surface, so that the thickness of the skin remains normal.

Starvation or semi-starvation also alters the composition of the body. Protein is not only not added to, but is used up, so that the cell mass of the body is reduced. Fat is used up, but the extracellular body fluid is actually increased. The loss of weight may thus be partly masked by what is known as famine oedema. The mechanism whereby this increase in body water occurs is not yet clearly understood.

As regards the detailed composition of a diet suitable for normal growth, much remains to be discovered. There must be an adequate supply of protein, which is the most important single item: nine different amino acids have been claimed to be essential for growth, and absence of any one will result in disordered or stunted growth. Protein malnutrition is a most serious problem throughout the world, and is a major factor in the condition known as kwashiorkor, in which there is a general slowing down of skeletal growth and maturation; the delay in epiphyseal fusion may be as much as a year compared with normally nourished children [p. 62], and puberty may be delayed. The muscles shrink, but subcutaneous fat may remain quite plentiful.

Other factors are also essential for growth. For example, zinc plays a part in protein synthesis and is a constituent of certain enzymes; a deficiency of zinc causes stunting, interference with sexual development, and falling out of hair. Iodine is needed for the manufacture of the thyroid hormones. Bone will not grow properly without an adequate supply of calcium, phosphorus, and other inorganic constituents such as magnesium and manganese. Iron is required for the production of haemoglobin, and fluorine is needed for the proper formation of tooth enamel as well as bone.

Vitamins play a most important part in growth. Vitamin A is thought to control the activities of both osteoblasts and osteoclasts. Too much vitamin A in the diet—as may occur when people take large amounts of vitamin preparations unnecessarily—can cause skeletal growth to slow down. On the other hand, a lack of vitamin A causes defects in the moulding process at the ends of the bones, so that clumsy thickened bones result, and the nerve trunks which run through holes in the skull may be pressed upon because the holes do not enlarge with growth as they should do.

Vitamin B_2 has a considerable general influence on growth, which is severely impaired in deficiency. In vitamin C deficiency the intercellular substance of bone is inadequately formed, and newly constructed cartilage lacks collagen. Gross deficiency causes scurvy, an important disease in which, among other things, there is a decreased growth rate at the epiphyseal plates, and deficient bone formation here and elsewhere, so that both the region of the plates and the shafts of the bones are easily broken; calcification is not interfered with.

In contrast, vitamin D deficiency is the cause of rickets, in which there is distorted growth in the region of the epiphyseal plates, as a result of deficient calcification of the cartilage there. This is because vitamin D stimulates the absorption of calcium from the intestine and possibly also the reabsorption of calcium by the kidney tabules. If there is too little vitamin D, there is therefore an inadequate supply of calcium and phosphorus in the blood stream, and in gross cases the softened bones may become distorted and bent under the weight of the body [PLATE 7]. Rickets is once again a significant disease in certain places in Britain, where the adolescent diet may be deficient in both vitamin D and calcium and where the children are unable to synthesize adequate amounts of vitamin D in their skin because of the lack of sunlight.

Finally, there is the question of oxygen. Children born with a congenital cardiac defect, of whatever sort, may, if the disturbance is severe enough, show a stunting and retardation of growth. If the defect can be corrected, for example by surgical repair of a cardiac anomaly, normal growth may result. The cause of the interference with growth is not wholly clear; it is not simply due to the tissues not receiving enough oxygen for normal metabolism, for the severity of the defect and the amount of interference with growth do not always run parallel. Furthermore, congenital cardiac defects which do not lead to deficient oxygenation of the blood may also be accompanied by impairment of growth.

SECULAR

There is an impressive series of figures to show that children nowadays are growing faster than formerly. The Norwegian Army measured its recruits from 1741 onwards, and found that in 1875 the average height was 1·3 cm more than it was in 1825 in recruits of the same age; by 1935 the average was over 5 cm more than in 1825. Similar, and perhaps more homogeneous, figures from Marlborough College in England showed that the average height of 16-year-old boys increased by more than 1·3 cm every ten years from 1873 to 1943.

A similar trend is found in younger children. Between 1880 and 1950 the average height of American and West European children between the ages of 5 and 7 years increased by more than 1·3 cm every ten years for a total of more than 10 cm. German, Polish, and Danish figures all confirm the general result, and interestingly enough it has been found that the increase is not confined to the less well off children. It follows that the phenomenon is probably not simply due to better nutrition in terms of calories, for one would expect an improvement due to this cause to be more marked in the poorer people than in the wealthy, who would presumably have been getting a good diet all the time. It is possible that the balance rather than the quantity of the diet may be a factor. There is some evidence that the trend is at last flattening off—at least in the United States—and it may be that the children of the well-to-do in Western Europe and America are now receiving a diet which is maximally efficient in promoting growth.

Although children are now growing faster, it seems that they are also stopping growing sooner. Early in this century most men did not reach their final height until about 25 years of age; since then the age has apparently dropped to 20 or less. It follows that the secular difference in height in adults is not nearly so marked as it is in children: adults *are* probably getting larger, but much more slowly than children of a given age. A rough rule for pre-school children is that the average secular gain in size has been about 1·3 cm in height and 0·5 kg in weight for each decade; in adolescents the gain has been about 2·5 cm and 2·3 kg for each decade. It is much more difficult to derive a figure for adults, since the times of achieving maximum height are changing, and there is also the progressive loss in height which occurs from 'middle age' onwards to be considered. Cross-sectional data may thus be very misleading, and longitudinal data are as yet too scarce to enable any firm pronouncement to be made.

In Britain weight is increasing secularly more than height, so that the adult male population (at least) is becoming progressively heavier at a given age. The most recent figures show that young and middle-aged men today are probably about 7 kg heavier than men of similar age and height thirty years ago. In earlier studies obesity increased with age until at least 55 years, but it appears that it now increases until about 35, after which it flattens off.

One of the most striking of the features which indicate a secular trend in maturation corresponding to the secular trends in growth is a progressive advancement in the timing of the menarche [FIG. 53]. The figures for this trend cannot be extrapolated backwards, as they would indicate that in the Middle Ages people never menstruated until late in life, whereas the historical evidence available suggests that puberty occurred around the age of 14 in mediaeval Europe as opposed to about 13 nowadays. The extrapolation of the curve forwards is equally alarming, as it would indicate that in a few years' time girls may be menstruating as babies. This is also an improbable inference, and it has been

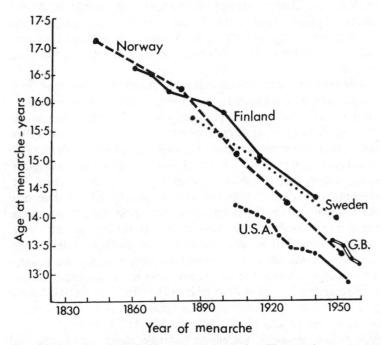

FIG. 53. Secular trends in the age of menarche. The graphs are based largely on interrogation of schoolchildren and young adults, but some of the early figures depend on the memories of older people. (From Tanner, J. M. (1962) *Growth at Adolescence*, 2nd ed., Oxford, Blackwell Scientific Publications, by kind permission of the author and publishers.)

suggested that what is now being observed is a recovery from a significant retardation of the menarche during the time of the industrial revolution. If this is true, the problem is to explain the cause of the hypothetical retardation. The revolution itself cannot have been an all-important factor, since figures from Scandinavia, where the revolution occurred much later and was much less violent, show an exactly similar trend. It has been put forward that the phenomenon could be due to a rise in world temperature, but climate has little effect on the menarche, as is shown by the fact that figures from Burma are almost exactly the same as those from Europe. A third, and more probable, suggestion is the progressive increase in the adequacy of nutrition in Europe and elsewhere. Nevertheless, the same reservations must be made in respect of this theory as were made in relation to the secular changes in height and weight. Recent evidence in Britain, North-West Europe and the United States suggests that the advancement of the menarche has stopped, but elsewhere in Europe it seems to be continuing.

Clearly, if epiphyseal fusion results from the secretion of hormones activated at puberty, it follows that the earlier puberty occurs the sooner the individual stops growing, and this might lead one to expect a secular *decrease*, rather than an increase, in the height of adults. The explanation of why this is not occurring probably lies in the fact that the children in whom the earlier puberty occurs are already taller [p. 152] than their predecessors, and this factor outweighs the earlier cessation of growth.

There is supposed to be a secular trend in the development of the intelligence quotient [p.113], but the difficulties in investigating this subject are formidable. The evidence suggests that there is an increasing superiority of girls over boys as regards I.Q. at the age of 11 years, and this has been related to the evidence indicating that the adolescent spurt in physical growth is occurring earlier.

SOCIO-ECONOMIC CLASS

Children in the top socio-economic group are taller than the children of unskilled labourers by about 2·5 cm at 3 years of age and about 4·5 cm at adolescence—a difference equivalent to twenty years secular trend. The difference in weight is less, but in the same direction.

This is not a simple matter of food, and probably several causes are involved. The economic factor seems to be less important than the provision of a home where meals are regular, adequate, and balanced, where sleep and exercise are sufficient, and where the children are taught the basic rules of health. The size of the family also appears to be important, for children in large families have been shown to be usually smaller and lighter than children in small families. Possibly this is because in large families children tend to get less individual care and attention.

These factors, together with the others considered in this chapter, emphasize the caution (already stressed) with which growth charts must be approached; charts and curves derived from children of a high social class in an economically advanced country may well be quite different from those taken from children of a lower class in an economically backward country.

SEASON AND CLIMATE

Growth in height is faster in the spring than in the autumn by a factor of 2–2½ times [FIG. 54]. On the contrary, growth in weight

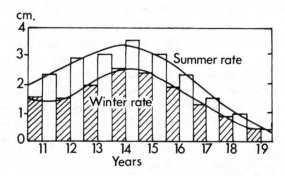

FIG. 54. Effects of season on growth. Half-yearly increments of growth in German cadets, 1902; growth in the summer months is more rapid than in the winter. Notice that the difference between summer and winter is greatest about the time of the adolescent spurt, and becomes less noticeable as maturity is neared. (From Thompson, D'Arcy W. (1942) *Growth and Form*, 2nd ed., London, Cambridge University Press, by kind permission of the publishers.)

proceeds faster (sometimes four or five times faster) in the autumn than in the spring. Children thus tend to grow in height in the early part of the year and to fill out in the latter part, though individuals may show a rhythm different from that of the majority. The reason why a seasonal difference in growth exists is not yet known, but it may have a hormonal basis.

It is suspected that the amount of daylight, which is a potent stimulus in other ways in the animal world, is a factor. Totally blind children do have seasonal variations in growth, but their fastest and slowest growth periods may occur in any season.

The seasonal tides of growth make it essential to measure any patient suspected of a growth disturbance over a full year [p. 30], for in autumn and winter growth in the height of children between the ages of 5 and 9 years may be less than the measurement error [p. 23]. At other ages the seasonal variations may be outweighed by overall changes of much greater magnitude.

Perhaps the most startling example of seasonal growth in animals is provided by the antlers of deer, in which every tissue concerned grows at a rate much in excess of its normal potential. Thus, the branches of the trigeminal nerve supplying the antlers may grow at a rate of as much as 12 mm a day [cf. p. 181]

Whether or not climate has a direct effect on growth is still doubtful; the effects formerly attributed to climate are difficult to separate from racial, dietary, and other effects such as those produced by disease. It has been suggested that each major race of mankind varies in stature according to the climates in which they live. Thus, the tallest whites live where winters are cold and wet, as in Scandinavia and northern Europe generally; on the other hand, Australian aborigines grow tallest under tropical monsoon conditions. It is safest for the time being to assume that, although there may be a geographical effect on height and weight, much of this is due to factors other than the meteorological conditions.

EXERCISE

It has already been mentioned [p. 55] that adults can increase the size of their muscles by exercise. The same holds good in children, though when the body is growing longitudinally the result is less marked than it is later, when the epiphyses have closed. As in adults, there appears to be no increase in the number of muscle fibres.

Exercise is also capable of reducing the storage depots of fat, and by so doing may alter the shape of the body and its composition. Inactivity is an important cause of obesity in children.

DISEASE

The effects of disease in childhood are similar to the effects of malnutrition, and good examples are afforded by tuberculosis and kidney disease. The administration of certain drugs—for example, steroids—may also retard growth. After disease or drug therapy ends there is an acceleration very comparable to that following the end of a time of malnutrition [p. 150], and again this homoeorrhesis may or may not completely compensate for the loss of growth during the period of illness. It is interesting that girls seem to be more resistant to the effects of outside influences interfering with growth than are boys, and after the influence has ceased to operate they 'catch up' with their normal curve more rapidly than boys do.

Maturation times in the female are also less affected by extraneous factors such as health and environment. In both sexes there is a homoeorrhetic spurt in maturation following illness, and this keeps pace with the spurt in growth.

Recent experimental work on piglets has shown that if they are delivered by Caesarian section under aseptic conditions and reared with minimal opportunities of acquiring any infectious disease, they mature faster: it is probable that similar considerations apply to humans.

Just how general diseases cause slowing of growth is not known, but it is suggested that there is a diminution in the secretion of somatotrophin, resulting from an increase in the secretion of corticosteroids from the suprarenal cortex. At all events, what seems to happen is that there is a decrease in the mitotic index of the cartilage cells in the epiphyseal plate, and that this, for a time at least, is not accompanied by a decrease in the rate of ossification. The epiphyseal plate therefore becomes thinner, and the number of cells available to produce more cartilage becomes reduced.

ANTENATAL INSULTS

There is some evidence that smoking by the mother during pregnancy, which is already known to cause a lower birth weight [p. 33] and an increased mortality round about the time of birth, can also influence later growth and development. In a large scale controlled investigation [p. 113] the 7 year old children of mothers who had smoked more than 10 cigarettes a day during pregnancy were on the average 1 cm shorter and 3–5 months retarded in reading ability when compared with children of non-smoking mothers. The differences are small, but they raise the possibility that events occurring before birth can produce changes in post-natal growth and maturation.

This proposition is supported by the recent observation that 'small for dates' full-term babies whose intra-uterine growth was slow are likely to continue to grow and develop slowly after birth. It has also been shown that undernutrition at a critical period of maximum growth velocity can cause permanent stunting of experimental animals, and this is associated with a deficiency in the numbers of cells in the organs of the animals.

EMOTION

A well-known experiment in post-war Germany goes to show the importance of emotional factors in growth. Two similar orphanages were selected for a dietary experiment. After a control period during which both institutions had the same diet, the children in the first orphanage were given a supplement to their diet. To the surprise of the experimenter, the growth of the children given the supplement actually fell behind that of the children in the other orphanage who were given no supplement. Investigation showed that the result was due to the transfer, at the same time as the control period ended, of an unpleasant superintendent from the second orphanage to the first. This female martinet used to take the children to task at the time when they were eating, and the severe strain imposed in this manner apparently outweighed the importance of the dietary supplement. The matter was clinched when it was found that she had transferred with her some of her 'favourites' from the second orphanage: these children had thrived under the original conditions, and grew even faster when given the supplement to the diet.

Emotional disturbances are a not uncommon cause of failure to grow normally, and the victims often show a bizarre and sometimes excessive appetite. They have been described as 'trash-can raiders'. It is not known how emotion influences growth, but there are several possibilities. It may be that they affect the hypothalamic growth centre, or that they disturb metabolism, or that increased amounts of steroids are produced. There may be associated dietary deficiencies in domestic cases of 'psychosocial dwarfism'. Recently it has been suggested that such children sleep poorly, and may not get enough stage 4 sleep; if so, the somatotrophin production would be diminished.

8

Growth and Repair

'What wound did ever heal but by degrees?'

SHAKESPEARE

GROWTH CYCLES: WEAR AND TEAR

All parts of the body in contact with the external environment require continuous replacement because they are necessarily exposed to repeated attrition and minor injury. Growth of this kind is not restricted to childhood, but continues throughout the whole of adult life. The most obvious example with which to begin is afforded by the skin and its appendages.

The superficial layer of the epidermis is constantly being worn away and shed, but the epidermis is maintained at a steady thickness by means of continuous growth in the deepest layer, in which lie the stem cells which replace the loss. The daughter cells produced by this 'germinative' layer [FIG. 55] are gradually pushed towards the surface by the steady division of the stem cells, and arrive there after about fifteen to twenty days in the case of the forearm skin. During their journey they lose their nuclei and their cytoplasm is progressively converted into keratin, a protein which serves to protect the deeper layers, but which, by its accumulation, leads to the death of the cells in which it is deposited.

Activity in the germinative layer of the epidermis can be detected by the appearance of mitotic figures. As might be expected, the skin does not grow all over the body at a constant rate, and regions which are subject to much friction exhibit more mitotic figures than regions which are more protected. There is also a diurnal rhythm of activity; for example, the cells in the

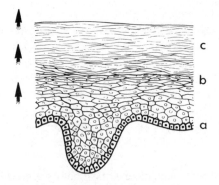

Fig. 55. Growth of epidermis.

(a) Layer of stem cells (germinative layer) in the depths of the epidermis.

(b) As the cells produced by the germinative layer are pushed towards the surface (and so further from their blood supply), they become flattened, lose their nuclei and die.

(c) The remains of the cells lose their identity and become converted into layers of keratin (cornified layer). Ultimately flakes of keratin are lost from the surface of the skin by attrition.

prepuce of an infant show more figures between the hours of 9 p.m. and 10 p.m. than between 5 a.m. and 10 a.m. It is a general finding, in animals as well as man, that the mitotic rate is high during rest or sleep periods, and low during wakefulness or activity. This has been correlated with the levels of adrenaline and corticosteroids in the blood. These are high during activity and low during rest, and it has been suggested that all diurnal rhythms of this kind are in part controlled by the circulating 'stress' hormones of the suprarenal, which are supposed to interact with the local chalone regulating mechanisms [p. 14] to suppress mitosis.

The growth of the skin adjusts itself to the varying surface area of the adult body. If there is a rapid and considerable gain in weight, the skin grows to cover the increased volume. Sometimes the volume may increase so rapidly that the skin cannot keep pace; for example, in pregnancy some areas of skin on the abdominal wall may become stretched and thinned, so giving rise to the characteristic thin shiny patches known as striae gravidarum.

When weight is lost in adult life the elasticity of the skin adjusts it to the lessened surface area, and skin growth becomes correspondingly depressed to keep pace with the needs of the body. If the loss of weight occurs after the elastic fibres have degenerated in old age [p. 222], the stretched skin hangs in loose wrinkles over the reduced body surface.

The cycle of growth, degeneration, loss, and repair characteristic of normal skin applies also to the hairs covering the body, which are constantly shed and replaced. The long hairs of the scalp have a life of several years, and the shorter hairs of the body may last for a year or so; the cycle is thus a much longer one than that of the epidermal cells. In all cases hairs tend to be replaced by other hairs, although in later life replacement is not complete [p. 216]. The root of the old hair is absorbed in a manner which is not completely understood, and the detached hair shaft is pushed towards the surface by the activity of the cells forming the sheath of the hair, until eventually it falls out. A new hair then produced by the division of cells in the 'matrix' of the follicle [FIG. 56], and grows up the follicle until it reaches the surface.

In every hair the period of growth is followed by a quiescent phase which ends in death. In the rat, all the body hairs show the same total life cycle, which is about 35 days, half being growth

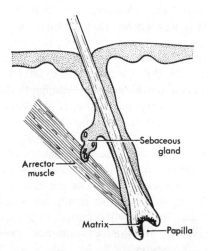

FIG. 56. Growth of hair. The cells of the matrix surrounding the papilla of the hair divide continually, so adding to the length of the hair and forcing it up its tubular follicle.

and half quiescence. Because hair replacement occurs in 'waves', it is relatively simple to determine the lifetime of rat hairs. In man, replacement occurs in scattered groups, so that some hairs in a small area of skin may be actively growing, while others are inactive or dying. In consequence it is not easy to tell whether or not there is a standard life cycle which applies to every hair.

The rate of hair growth varies with the texture of the hair: coarse hair grows faster than fine hair. It also appears that hair grows more rapidly in summer than in winter, but it is quite unaffected by repeated cutting, or by exposure to sunshine. Every day something between 20 and 75 scalp hairs are moulted, and the scalp hair grows at the rate of about 2·1 mm a week; the corresponding rate for the beard is higher, perhaps 2·8 mm. The hair on the legs of children grows at a rate of 1·4 mm a week; this increases to nearly 2 mm a week by the age of 40 to 45 years. The axillary hairs grow 2·2 mm a week, and this rate is unaffected by age. Hair may continue to grow at its normal rate even though other tissues in the body are severely retarded by malnutrition.

There is a hereditary tendency towards the development of baldness in males, and it is interesting that castration appears to keep this in check. The administration of male sex hormone does not in itself produce baldness, but if it is given to castrates it will permit the genetic factor to operate, so that the individual will become bald. It is well known that baldness of this type is very unusual in women.

The growth of nails differs in some important respects from that of the hairs. In the first place, it is affected by nutrition. Secondly, there is good evidence that damage to the nails, such as is produced by nail biting, leads to a 20 per cent higher rate of growth. The finger-nails grow about four times as fast as the toe-nails, and this effect is not yet explained, although a contributory factor may be that the finger-nails are trimmed more frequently than the toe-nails. The average growth of the thumb-nail is about 4 cm a year, and the rate appears to depend in part on the amount of usage of the hand, for when the hand is immobilized because of injury the growth of the nails is considerably retarded. Finally, like the hairs, the nails grow faster in the summer than in the winter.

Less is known about the repair and replacement of other surface structures. The epithelia of the respiratory and urinary apparatus renew themselves comparatively slowly, but those of the gastrointestinal tract are renewed at an almost frightening rate. The new cells are formed in the crypts of the intestine and migrate along the villi to their tips, from which they are shed into the lumen of the gut. It is estimated that in the rat the mucosa of the small intestine is completely replaced every 34 hours, and it is probable that the cycle is similar in length in man; mitotic figures are very common in the human intestine. The scale of this phenomenon is not often appreciated. It has been calculated that a rat loses daily about one-twentieth of the total cell population of its whole body.

In vitamin A deficiency certain epithelial surfaces fail to shed their dead cells and in consequence become dry and thickened. In particular the cornea may soften and ultimately disintegrate, causing blindness.

The source of the injury responsible for cyclical death and replacement is not necessarily an external one. The red blood cells which circulate round the body have to squeeze through narrow channels during the whole of their life in the vascular system; at the same time they have no nucleus which might enable them to undertake 'running repairs'. As a result they become worn out after a life of approximately 120 days, and their condition is in some way recognized by the cells of the reticulo-endothelial system, which seize and destroy them. The loss of circulating red cells is made good by activity of the stem cells in the red bone marrow, which in the adult is found in the vertebrae, the flat bones, and the ends of the long bones. The capacity of the marrow to cope with red cell destruction is indicated by the fact that the loss of a half a litre or so of blood by a blood donor is fully restored in the course of a few days. After repeated haemorrhages the red marrow is found to have invaded the shafts of the long bones, indicating the immense effort the body has been making to replace the loss. Should the marrow fail to keep up a balancing supply of new cells, the red cell count falls and the patient becomes anaemic.

The genital apparatus in both male and female affords examples of continual growth activity. In the testicle the spermatozoa are produced by the stem cells in the walls of the seminiferous tubules, and are gradually pushed out of the testicle by the pressure of their fellows. It is not known how long they live in the tubes of the male genital system, but each ejaculation contains up to 100 million spermatozoa, and their life outside the body is very short. In animals there is evidence that spermatogenesis takes place in 'waves' of activity which pass along the seminiferous tubules, and probably the same sort of thing takes place in man.

In the ovary there is a fairly regular cycle of growth which results in the production of one mature ovum (rarely two) every month or so. There is quite considerable variation in the exact timing, which may indeed differ in the same woman from cycle to cycle. Most women have a cyclical period which lies somewhere between 25 and 35 days, and a 28-day cycle is accepted as the average. One (or possibly two) of the primary oocytes which are present in the ovary at puberty becomes surrounded by a layer of cells which divide and grow and eventually surround a cavity

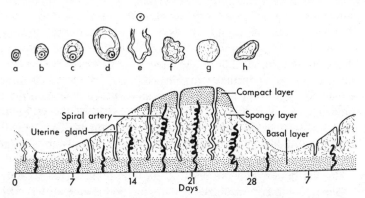

FIG. 57. Ovarian and menstrual cycles. The phases of the development of the ovum are shown above, while the phases of the uterine cycle corresponding to them are indicated below.

(a) Primary oocyte. Uterine mucosa being shed in menstruation.
(b) Development of follicle. Mucosa beginning to grow.
(c) Appearance of cavity in follicle. Uterine glands developing from stumps left in basal layer.
(d) Secondary oocyte surrounded by mature follicle. Glands enlarged but not yet tortuous: mucosa thickening.
(e) Follicle bursts, releasing secondary oocyte and its coverings. Compact layer of uterine mucosa developing.
(f) Follicle full of blood clot; corpus luteum forming. Uterine glands now tortuous and arteries becoming spiral.
(g) Corpus luteum of menstruation. Full development of uterine mucosa.
(h) Degenerating corpus luteum. Menstruation begins.

[FIG. 57]. During this time the primary oocyte undergoes the first stage of a reduction division which forms a secondary oocyte plus a polar body. Further growth in size of the whole apparatus, which is now known as a Graafian follicle, is followed by bursting of the follicle, so allowing the secondary oocyte and its coverings to be shed into the Fallopian tube. In the tube the second stage of division occurs when fertilization by the spermatozoon takes place, so forming the mature ovum and a second polar body.

The cavity from which the secondary oocyte was expelled becomes filled with blood clot, and the cells of its walls multiply to form a yellow structure which is called the corpus luteum of menstruation. If pregnancy does not supervene, this becomes converted into scar tissue [p. 173], and is then called a corpus albicans. If the ovum is fertilized, it embeds itself in the wall of the uterus and this causes the corpus luteum of menstruation to grow and become corpus luteum of pregnancy, an endocrine gland which secretes hormones until about the fifth or sixth month of pregnancy.

Corresponding with, and influenced by, the ovarian cycles are the cycles of growth, death, and shedding of the inner layers of the wall of the uterus which are responsible for menstruation, and are therefore called the menstrual cycle. These changes, like so many aspects of growth, are under the control of the endocrine system, in particular the ovary and the anterior lobe of the pituitary, and have the purpose of providing every month a fresh epithelium suitable for the reception of the fertilized ovum.

The uterine lining (endometrium) at the end of menstruation consists of a thin layer, in which are found the stumps of the uterine glands [FIG. 57]. Under the influence of oestrogens produced by the developing ovarian follicle, the epithelium of these glands rapidly regenerates and spreads over the whole surface of the uterine cavity by the end of the first week of the cycle. From the second week onwards there is spectacular mitotic activity in the endometrium, which thickens markedly; the glands enlarge, and the blood vessels proliferate. At the same time the endometrium becomes swollen with fluid retained under the influence of oestrogens.

After ovulation, the corpus luteum secretes progesterone, and under its action the glandular cells of the endometrium enlarge, accumulate glycogen and fatty material, and begin to secrete. The glands become tortuous and dilated, and the endometrium thickens still further, until it forms three recognizable layers. The innermost layer is relatively compact, the middle layer is filled by intensely tortuous 'corkscrew' glands and by large spirally-arranged arteries, so that it is called the spongy layer, and the basal layer remains much as it was at the beginning of the cycle. At the same time, the muscular wall of the uterus becomes engorged with blood and begins to exhibit rhythmic contractions.

Eventually, just before menstruation starts, changes in the balance of endocrine stimulation lead to a diminution of the extracellular fluid and the spiral arteries constrict. This deprives the superficial layers of the endometrium of their blood supply, and they are shed into the uterine cavity. The constriction of the arteries relaxes at intervals during menstruation, and thus contributes to the bleeding, which is probably mostly the result of the tearing across of veins.

This cycle has been described in some detail because it illustrates what a complicated process a growth cycle can be, and because it affords perhaps the best example of extremely rapid growth and repair occurring under normal conditions in the intact human body. A good deal is known about the menstrual cycle, but much remains to be discovered, in particular in relation to the hormonal control of the process.

It must be realized that growth cycles take place in virtually every tissue and system, with the notable exception of the central nervous system [p. 55]. The only difference between them and the ones described is that much less is known of the details, and that information is only now coming to hand. For example, it was formerly thought that cells which were highly specialized had little capacity to divide and grow. A good example was the liver, in which the cells have a specialized biochemical function, and it was believed that repair or replacement of these cells could not occur. However, it has now been shown that there is quite a considerable turnover of liver cells, and mitotic figures can be observed in the liver if they are carefully sought for, though the mitotic index [p. 10] is relatively low. Liver cells grow, divide, become old, and die, only to be replaced by new cells so that the size and weight of the organ show little alteration throughout adult life. The average life of a liver cell has been estimated at about eighteen months.

REGENERATION

When an injury greater than that imposed by normal wear and tear is suffered, the local cells are stimulated to divide until the injury is repaired as far as possible. The nature of the stimulus is quite unknown, although it is thought that the injured cells may produce chemicals which initiate the sequence of events concerned in healing. Since cells are better at dividing in young children than in adults, it is, theoretically at least, an advantage to be a child when repairing tissue damage.

The ideal arrangement would be for a complete 'de-differentiation' of the local tissue cells to take place, so that they reverted to their embryonic condition and could then reproduce all the various kinds of tissue which were destroyed by the injury. This is a power possessed by many lower animals, but it has been lost somewhere in the course of evolution. It is conceivable that medicine in the future may enable us to recover it.

Animals such as *Hydra* and flat worms can grow entire new bodies from a fragment of the old one. Star-fish can regenerate a limb from a body or a body from a limb. Lobsters can also regenerate a limb, though the process takes a long time, and the new claw is always smaller than the original one; each time a limb is lost, the regenerating power diminishes. Fish can grow new fins, and lizards new tails.

In the early stages of regeneration of a limb or tail phagocytic cells remove debris from the site, and large numbers of connective tissue cells and fibres appear, accompanied by blood capillaries. During this preliminary phase there is a destruction of protein and an excretion of nitrogen; this appears to be an essential part of the process whereby 'rejuvenated' and 'de-differentiated' cells are produced.

After about a week the connective tissue cells are replaced by an accumulation of relatively undifferentiated cells, derived from connective tissue, muscle, and cartilage in the vicinity. These cells are surrounded by intercellular matrix, and multiply rapidly, forming what is called a blastema. Later still, they differentiate again into muscle, cartilage, etc., and a new limb or tail grows in much the same way as it did in the embryo. During this second phase, protein is built up and an adequate amount of dietary protein is essential.

The whole of this process appears to require the presence of a certain density of sensory innervation, and the influence of intact nerve fibres is particularly important during the phase in which the blastema forms and becomes differentiated. The nerve fibres which grow out and invade the regenerating stump are very different from normal adult nerve fibres, and contain tubules which could conceivably serve to conduct to the site chemicals elaborated by the cell body [p. 143]. The importance of the local innervation is demonstrated by experiments on frogs. Normally an adult frog cannot grow a new limb, though a tadpole can. But if the front leg of a frog is amputated and the sciatic nerve is simultaneously transplanted from the hind limb into the forelimb stump, a new (admittedly imperfect) limb can be regenerated. It is the sensory innervation that is vital; if the sensory ganglion is destroyed before transplanting the nerve, the limb fails to regenerate.

The rate of regeneration, like the rate of growth, appears to decline with age, and shows the same variation with seasons [p. 156]. Vitamins, hormones, and other factors influencing growth [CHAPTER 7] have similar effects in relation to regeneration.

Regeneration of this kind is sometimes called 'major regeneration' on the assumption that it involves true de-differentiation of cells to a multipotent stage, so that a cell derived from one tissue can give rise to cells characteristic of another. This may be distinguished from 'minor regeneration', in which there is little or no evidence of such drastic preliminary de-differentiation, and in which each tissue has to make good its own losses by the mere stimulation of mitosis in cells which have retained the capacity to divide under appropriate stimulation. In man, major regeneration does not occur, but minor regeneration follows the infliction of a wound or other destructive process. What is possible can best be illustrated by concrete examples, and the most obvious one with which to begin is an injury to the skin.

REPAIR OF SKIN

When the skin is injured—for example, by the surgeon's knife—the region of the damage is immediately occupied by blood clot. Following this there is a delay, which probably corresponds to the initial delay phase of animal regeneration [p. 169]. This delay is a phase of essential preparation for subsequent events. In a short time scavenging macrophages appear at the margins of the clot, having migrated out from the intact connective tissue surrounding the injury. They invade the clot and begin removing it, increasing in size as they do so because of the amount of material they ingest.

Immediately following these scavenging cells are the first new blood vessels. These are formed from the lining cells of intact blood vessels at the margin of the wound, and resemble capillaries very closely. The little sprouts grow steadily into the clot, and they are at first solid; later they become canalized, join up with each other, and convey blood through the loops thus formed. Later still smooth muscle cells appear—from what source is not certain—and surround the largest of the developing vessels, which in this way are transformed into arterioles; they then acquire an outer coating of 'adventitial' cells, presumably derived from the local connective tissue. Last of all, nerve fibres invade the region and seek out the muscle cells of the new vessels, thereby providing them with vasomotor control.

A similar series of outgrowths occurs from the local lymphatic capillaries, but there appears to be no communication between the lymphatics and the blood capillaries, as was formerly supposed.

So far, then, the clot produced has been invaded, debris has been removed, and the newly forming tissue has been supplied with nutriment by blood vessels. The stage is now set for the production of the collagen fibres which will heal the wound and maintain its tensile strength. The exact process by which this occurs is still a matter of debate, but it is thought that the fibroblasts in the surrounding connective tissue are stimulated to divide and then to invade the clot about the same time as the blood vessels. This fibroblastic activity is perhaps the most striking feature of a healing wound. The production of collagen fibres [p. 0] takes time, and in experiments on the rabbit's ear it has been found that the first collagen fibres do not appear until about six days after the fibroblasts invade the clot. Most of the collagen is laid down during the second week of healing, and the tensile strength of the wound increases as the collagen fibres grow and increase in number. Elastic fibres are regenerated somewhat later than the collagen fibres, but the details of this process are almost wholly unknown.

Meanwhile, the epithelial cells at the side of the wound migrate sideways on to the surface of the clot and newly formed tissue. Some cells move down into the incision, but as healing proceeds these ingrowths are discouraged and the intrusive cells are removed. Sometimes epithelial cells may be driven down into the dermis at the time of infliction of the injury, and they may form small clumps of epidermal growth there which may give rise to an 'implantation cyst'. Epithelial cells are not normally motile, and what causes them to move across an injured surface has not yet been explained. It may be that a common chemical stimulus encourages both motility and the subsequent outbursts of mitosis among the epithelial cells.

Following injury, the mitosis rate in the epithelium surrounding the wound rises to a figure about twenty times as great as that for normal skin. Initially mitoses are found only in the normal tissue round the edges, but later the migrating cells begin to divide and mitotic figures can be seen on the surface of the injured area.

When the epithelium migrating from one side of the wound meets epithelium migrating from the other, migration ceases, and here again we do not know why, though it is commonly believed that the physical contact between adjacent epithelial cells is the stimulus. In the initial stages following the establishment of contact the layer of epithelium covering the clot and the underlying repair processes is very thin—only one cell thick—but later the new epithelium thickens up until a reasonably normal cover is obtained. However, structures such as hairs and sweat glands may not be completely replaced, and the new epithelium may lack the normal patterning of the skin, so that it looks shiny and glazed.

If the area of a skin wound is more extensive, what is called 'granulation tissue' develops. The base and edges of the wound swell from the accumulation of fluid and cells from the damaged blood vessels. Later, blood vessels invade the clot and debris, and the whole raw surface becomes red and granular in appearance. Each granulation has a core of new blood vessels which are very delicate and bleed easily. The new tissue increases in thickness until the defect is filled [Fig. 58]. Usually this process stops when the granulations reach the level of the surrounding skin, but occasionally they proceed to bulge beyond this level, producing

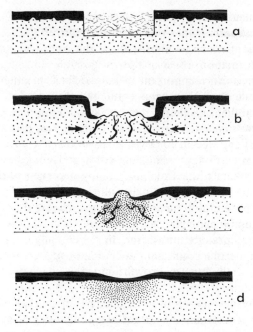

FIG. 58. Scheme showing healing of a granulating wound.

(a) Condition after infliction of wound, showing loss of tissue and blood and debris on the surface.

(b) Blood vessels have invaded the debris and granulations are formed. The arrows indicate invasion of the dermis by fibroblasts and also ingrowth of epidermis to cover the raw surface.

(c) Healing nearly complete. Scar tissue forming in the dermis.

(d) Complete epithelialization. Note that the epidermis is usually thinner and that its deep aspect is smoother than normal. Scar tissue in dermis.

what are called 'exuberant' granulations. Exactly similar events occur in granulation tissue as have already been described: invasion by capillaries is followed by invasion by fibroblasts, and eventually collagen is formed. Now it is a feature of newly formed collagenous tissue that it tends to contract, though whether it is the collagen fibres themselves or their 'parent' fibroblasts which do the contracting is uncertain. The process is of the greatest importance in healing, since it pulls the edges of the wound together, and the collagen gives the resulting scar its strength.

In the early stages the collagen can be stretched by tensile forces applied to the wound, and recent animal work on the healing of experimental wounds has emphasized the fact that very much more needs to be known about the contraction of collagen and the formation of scars. It is possible that this kind of mechanism is more important in animal than in human skin.

Meanwhile there has been epithelial regeneration of exactly the same kind as in a simple incised wound. However, the process is naturally very much slower if epithelium has to come in from the sides of the wound to cover a very extensive defect, and the wound is much longer in healing completely. If some epithelial structures are left intact or only slightly damaged in the base of the wound—for example, hair follicles or sweat glands—the epithelia of these structures will also be stimulated to divide, and little islands of epithelialization will be formed from which the surrounding granulation tissue is eventually covered. The epithelium becomes adherent to the growing granulations, and this stops their further growth, but the mechanism is again unknown.

This brief description of what happens in a simple skin wound may serve to indicate the extent of our ignorance regarding the simplest problems of growth and repair. Nevertheless, something is being discovered about the factors which affect wound healing, and this is of great importance in clinical medicine and surgery. For a long time it has been suspected that a chemical stimulus [p. 5] could influence the invasion and proliferation of the repair organization, and clearly if this stimulus could be identified we might have a powerful weapon to use in surgery. Recent investigations suggest that the 'stimulus' may in fact be simply the loss from the wound of the tissue chalones [p. 14] which inhibit mitosis. We know that a good blood supply is necessary for healing to proceed normally, and it has often been said that wounds of the face heal more quickly than wounds of (say) the foot because the face has such an excellent blood supply. Experimental evidence for this belief is lacking, as is evidence to support the other common statement that wounds of areas exposed frequently to injury heal more quickly than wounds of more protected areas of skin.

The presence of infection in the injured tissues, or the inclusion of irritant substances in the wound area may delay healing greatly, and if there is a general disease present, such as diabetes, the ability of the tissues to respond to the emergency may be impaired.

Although it is certain that age is a factor in cell division, wounds in old age heal reasonably well provided that the blood supply is good and there is no disease of blood vessels. A good balanced diet is desirable, but minor protein deficiency has comparatively little effect, for protein seems to be diverted preferentially to the healing wound. Severe protein deficiency causes delay in healing. A lack of vitamin C has little noticeable effect on the regrowth of epithelium, but the formation of collagen is rendered deficient, and wounds may break down because of their poor tensile strength.

Zinc is needed for the normal healing of wounds. Concentrations of zinc increase preferentially at the site of a healing wound, and zinc sulphate added to the diet of experimental animals promotes wound healing, although no histological reason for this observation can be adduced.

A curious observation made in rabbits after the infliction of a small whole-thickness punched wound in the ear is that tissue regeneration in such circumstances is faster in males than in females. It has been shown that the effect is general rather than local, and that treatment with testosterone stimulates regeneration in the female. Conversely, removal of the testes from a male rabbit is followed by a decreased rate of regeneration, provided the operation is carried out some time before the infliction of the wound.

Finally, hormones derived from the suprarenal and pituitary glands can be shown to influence the healing of experimental wounds in animals, but it is questionable whether this can be of much importance in man, since the disturbance has to be quite severe in order to produce an effect.

REPAIR OF FRACTURES

When a bone is broken, just as when skin is damaged, there is a zone of haemorrhage at the site of injury. This is penetrated by macrophages, and later by fibroblasts, which lay down collagen in the gap between the broken ends, so forming a scaffolding on which calcium salts are deposited to form what is called provisional callus. At the same time the bone-forming layer of the periosteum, which the injury may have displaced some little distance away from the shaft of the bone, starts to form new bone in the tissues surrounding the break [PLATE 8]. While this is going on, other cells are producing matrix within which stem cells from the periosteum differentiate into either cartilage-forming or bone-forming cells. In some instances cartilage is formed in large amounts, but in others bone formation predominates from the beginning. In either case, the cartilage is ultimately converted into bone [p. 58].

In the initial stages, bone is laid down more or less at random in the vicinity of the break, but later on, moulding of the kind described earlier [p. 65] takes place, and the stresses and strains around the fracture cause the trabeculae to be laid down in a definite pattern. Unwanted bone is then removed by osteoclasts brought to the site by the blood vessels, and eventually a more streamlined structure is formed, which may restore the original line of the bone extremely well [FIG. 59]. This whole process may take months or even years, but the resulting healing of the fracture produces a structure adapted to resist the new forces which may be brought to bear on the bone if the fragments do

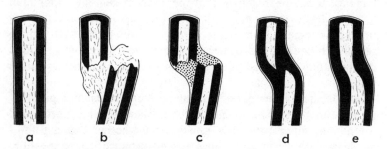

a   b   c   d   e

FIG. 59. Healing of fracture.

(a) Normal long bone.
(b) Situation immediately after break. There is displacement and angulation of the fragments; the periosteum has been torn and displaced, and there is haemorrhage between and around the fragments.
(c) The torn ends of the periosteum have united, and the osteogenic layer is laying down bone in the region outside the break (external callus). At the same time osteoblasts are producing callus in the region of the marrow cavity (internal callus).
(d) The broken ends are united by a clumsy but efficient bridge of bone.
(e) The unnecessary new bone has been removed. In the final arrangement the architecture of the bone depends on the stresses to which the region is subject as a result of the altered mechanics of the healed bone.

not unite in precisely the same alignment as they had before the break.

In a child the correction of a badly set fracture of the wrist—even one with a malalignment of as much as 60 degrees—may be complete in 3 months, and this raises the question of the control of such healing, for it implies the existence of a 'model' of the proper shape for each part of the body; the part must be restored to correspond as nearly as possible to this model. Similarly, in attempting to straighten out a club foot the surgeon finds that the foot is always 'trying' to return to its distorted shape, and it has been suggested that in such cases the 'model' is a distorted one. No one yet knows whether such 'models' are the expression of some process in the central nervous system, or whether they are inherent in every cell of the body.

The healing of a fracture, like that of a skin wound, requires vitamin C and protein, but in addition depends on adequate supplies of mineral salts such as calcium phosphate, normal amounts of parathyroid hormone, and probably many other factors. Vitamin deficiencies tend to produce characteristic differences in the healing process.

REPAIR OF OTHER TISSUES AND ORGANS

Nerve cells are perhaps the most specialized cells in the body, and they cannot divide to replace themselves. Following injury to the central nervous system attempts are made at regeneration by the supporting connective tissue cells, which behave in much the same way as fibroblasts do in the repair of injuries elsewhere. The damaged region is thus ultimately replaced by scar tissue, and there is no replacement of nerve cells which have been destroyed. Functional deficiencies produced by cell destruction in the central nervous system are thus permanent.

Human striated muscle is much less capable of regeneration than the muscle of amphibians, which has been more intensively studied. Still, regeneration can make good a certain amount of destruction of muscle in man. The dead tissue is removed as usual by the action of macrophages, and the new fibres are derived from the satellite cells at the margins of the dead area. It appears to be important that the connective tissue tubes which acted as sheaths for the original fibres should remain intact, so that the new fibres can grow along these tubes. The main factor which hinders regeneration of striated muscle in man is the relative ease with which scar tissue formed by collagen can be manufactured. Connective tissue cells are, so to speak, much better at repair work than satellite cells, and scar tissue forms in between the specialized cells, obstructing their progress, and closing the tubes off, before the satellite cells have had time to produce new muscle cells.

Cardiac muscle seems to have virtually no capacity for repair and regeneration, and any portion of the heart wall which dies following a blockage of its blood supply is repaired by fibrous tissue.

Regeneration of tendon can be reasonably satisfactory, but full functional recovery of a completely severed tendon is unusual, since the cut ends tend to unite, not only with each other, but also with the local connective tissue, so limiting movement.

Articular cartilage which has been damaged may become converted into fibrous tissue. But if it is detached from the end of the bone and can find a protected region in the synovial cavity it may survive and even grow, being nourished by the synovial fluid. The knee joint, with its liability to injury and its extensive synovial cavity, is particularly prone to harbour such pieces of cartilage.

Small blood and lymph vessels are capable of complete and rapid repair, and a damaged lymph node can reconstruct itself, though if one is completely removed it is not replaced except in young animals.

Regeneration of internal organs such as the pancreas, the salivary glands, and the endocrine glands is fairly limited, though it is possible that given certain conditions these organs can replace themselves better than is at present supposed. Although renal epithelium can repair itself fairly well, the kidney has to compensate for loss of kidney tissue by hypertrophy [p. 200] of the remaining tissue as well as by cellular replacement. On the other hand, the urinary bladder is remarkably efficient as regards regeneration. The lungs cannot regenerate if damaged, though lung tissue will expand to fill up the cavity left by removal of a lobe or segment.

In many cases the response of hypertrophy can be shown to be dependent on functional demand. Thus, although a congenital absence of one kidney may be accompanied by hypertrophy of the other after birth, such hypertrophy is not evident in cases of stillbirth where the kidney has never worked. Again, hypertrophy of the thyroid gland following partial removal can be prevented by the administration of thyroid extract.

The internal organ which has been subjected to most investigation is the liver, for it can regenerate to a quite extraordinary extent following injury or damage produced by disease. Most of this work has been done on animals, but now that advances in surgery have made it possible to take out part of the human liver in the treatment of disease and injury, more and more importance is being attached to human studies. It does not seem to matter which part of the liver is removed; regeneration proceeds equally well from the remaining material, and the lobular structure of the liver is preserved in the newly formed mass. The maximal rate of liver regeneration in animals is not far short of the rate of growth in the embryo of the same species, and it has been suggested that in the first 10 days of regeneration following partial removal of the human liver as much as 100 g of liver tissue may be added each day; after the first two postoperative weeks there is little change in the size of the liver remnant.

The control of liver regeneration raises many questions. For example, it is known that the body can get along quite well with only a fraction of the normal amount of liver tissue. If so, why should the liver start manufacturing new cells when there is no immediate need for them? In young actively growing rats the restoration of the liver mass may actually exceed the original provision, often by as much as 50 per cent.

Again, what is the stimulus which causes regeneration to take place? And how does the body know when sufficient liver tissue has been formed, so that regeneration comes to a halt? These questions are similar to those already asked regarding the repair of a skin injury, and they have as yet no answers.

Nor is it known whether the goal of the repair process is the restoration of a critical morphological mass or the restoration of a reserve of physiological function. In the embryo, growth of organs occurs even though they have not yet begun to function; presumably this phase of growth is under genetic control, and is not mediated by physiological requirements. After birth the situation seems to change. In some tissues and organs, such as the liver, the blood cells, and possibly the ovary, complete restoration of the original mass can occur, whereas in others, such as somatic muscle and nervous tissue, almost no attempt at specialized repair is made. To account for these findings a feed-back mechanism based on functional demand has been postulated. Attempts have been made to find out whether artificial restoration of tissue mass influences repair. For example, following partial removal of the liver the original mass of liver cells has been made up by injecting homogenates; the results have not been clear cut, and are often conflicting. Transfusion of red cells depresses haemopoiesis in patients with anaemia, but this may be due to factors other than the restoration of the red cell mass.

Another interesting problem is presented by the difference between the repair of a local injury such as a skin wound, in which the activity is also strictly local, and the repair processes which occur in organs like the liver, in which the response is not purely local, but affects distant parts of the whole organ. More striking still is the condition of compensatory hypertrophy [p. 200] following removal of one of a pair of organs, such as the kidney. In such cases the enlargement of the remaining organ requires the postulation of some generally distributed influence which affects only the 'target' organ, since generalized overgrowth of other tissues does not occur. It is difficult to escape the conclusion that this influence is the functional stress placed on the surviving member of the pair.

REGENERATION OF PERIPHERAL NERVE

The lack of functional repair in the central nervous system must be contrasted with the results of injury to the peripheral nerves, which are composed of the processes of cells which have their cell bodies in or near the central nervous system. The injury in this case does not involve the nuclei of the damaged cells, and there is little difficulty in replacing the cytoplasm which has been destroyed, just as an amoeba can make up for the loss of a pseudopodium. Indeed, the cell body is, quite apart from any injury, constantly synthesizing protoplasm which passes slowly down the various processes of the cell to make good the wear and tear on the protein systems they contain.

After section of a peripheral nerve, the portions of the individual nerve fibres that have been cut off from their parent nerve cells split up into small fragments, die, and are removed by macrophages which penetrate the nerve from the local connective tissue. At the same time the myelin sheaths of the myelinated fibres in the nerve disintegrate into fatty globules, which are similarly removed. The result is that the nerve beyond the point of section comes to consist of a series of tubes which are cleaned out ready for the regeneration of the fibres to take place. This it does at the rate of about 1–3 millimetres a day, material synthesized in the cell body travelling down the fibre towards the periphery. This may sound quite fast, but it means that after a section of the sciatic nerve in the buttock it may be anything up to three years before the growing fibres reach the foot. Even if nothing goes wrong, the repair of a nerve injury is a slow process.

The injury to the nerve produces the usual haemorrhage between the cut ends of the nerve trunk, and this clot is penetrated and organized in the usual manner by macrophages, blood vessels and fibroblasts. Accordingly, when the regenerating nerve fibres try to find their way across the gap they are confronted by a dense thicket of cells, collagen, and matrix which it is very difficult for them to penetrate. Those which do, enter a tube in the farther stump which may or may not be the one which they formerly occupied. (Since there may be several thousand fibres in the nerve, the likelihood of any one fibre reaching its former tube is pretty small.) Other fibres do not succeed in getting through the obstacle, and double back on themselves, or else skirt round the scar tissue, being guided into the peripheral stump, if they reach it at all, by what appears to be a chemical stimulus.

The result of all this is that many fibres do succeed in getting into the peripheral stump, but that very few of these will be conducted by the empty tubes to their proper destination [FIG. 60]. Should a motor nerve fibre reach a sensory endorgan, or vice versa, it degenerates and disappears. A fibre which reaches an appropriate destination is at first much thinner than normal, but when it establishes a connexion with a suitable endorgan the fibre thickens and the fibre/endorgan complex begins to function. The probability is very great that the endorgan reached will not be the same as the one originally supplied by that particular fibre, and a complicated programme of re-education in the central nervous system is therefore necessary. For example, if the patient wishes to move the little finger after section of the ulnar nerve, and if

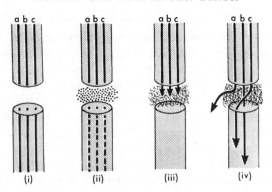

FIG. 60. Regeneration of nerve fibres.

(1) Situation just after section of the nerve. Three of the cut fibres in it are indicated diagrammatically.

(ii) Blood clot and debris occupy the space between the cut ends and the peripheral parts of the fibres are degenerating.

(iii) The peripheral stump has been cleared of the remains of the degenerated fibres and is ready to receive those which are beginning to regenerate. The clot has become converted into fibrous tissue.

(v) The fate of the three fibres. A has failed to penetrate the connective tissue barrier and is growing into the tissues outside the peripheral stump altogether. B has succeeded in growing along its own peripheral tube (a rare event) and will establish its former connexions once again. C has found its way into the peripheral stump, but has become diverted into a tube other than its own.

several of the fibres which originally helped him to do so have found a new destination in one of the small muscles of the thumb, it is disconcerting for him to find that his willed movement of the little finger actually causes contraction of the thumb muscle. Re-education can do something to overcome this, but the phenomenon is always troublesome, and good functional results can only be expected in little children, who have not yet established rigorous central patterns of movement.

While the regeneration process is under way, other curious events take place in the periphery. Many nerves which supply skin have an 'autonomous zone' which is supplied by that particular nerve and by no other. Around this central area lie the 'overlap zones', which are supplied by the nerve in question, and also by one or more of the adjacent nerves [FIG. 61]. When a cutaneous nerve is cut, the result is therefore an area of anaesthesia surrounded by a region in which partial sensation persists,

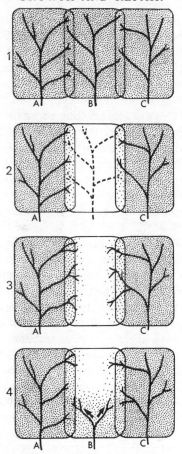

Fig. 61. Reinnervation of skin.

1. Patch of skin showing normal overlap of three nerve territories (other overlapping territories are omitted for the sake of clarity).

2. When the nerve B is cut, sensation is completely lost in its autonomous zone, but in the overlap zone a degree of sensation persists due to the intact overlapping nerve terminals.

3. Following degeneration of the nerve, the area of complete sensory loss shrinks, because sprouts from nerves A and C invade the autonomous zone of B. Sensation in the overlap zone also improves.

4. When the cut nerve regenerates, these sprouts regress in proportion as the normal anatomy is restored. If adequate recovery of nerve B does not occur, the sprouts persist, and the condition in diagram (3) becomes permanent. The amount of ingrowth is limited, and, unless the denervated area is very small, can never make up completely for the loss.

because some of the fibres supplying this zone are still active. It is found that the zone of partial anaesthesia begins to shrink shortly after the injury, because there is an ingrowth of sprouts from the peripheral ends of the adjacent intact nerve fibres. It appears that some stimulus arising in the denervated area causes them to grow in to try to make good the deficiency.

When the fibres of the regenerating nerve arrive at the denervated zone, the invaders from the periphery tend to degenerate and be removed. One could imagine that there was some mathematical formula regarding the density of innervation desirable in a given area of skin. If this density is reduced, fibres grow in to try to make it up again; when the desirable figure is exceeded, excess fibres are removed. How this process is controlled we have no idea.

Nerve fibres will invade a skin graft in a similar manner, and eventually the graft will attain a certain measure of sensitivity. Scar tissue is also invaded by sensory fibres, but because of the density of the tissue it is difficult for them to make their way so easily through it as they do in more or less normal skin. Innervation of scar tissue is therefore never completed up to 'normal' density, and the nerve fibres tend to be isolated from each other.

Experimental work has shown that nerve fibres can be extended far beyond their original length by deliberately misdirecting them. Fibres from one nerve can be induced to grow into the degenerated peripheral stump of another, and once regeneration has reached a certain stage this nerve can in turn be cut and introduced into the stump of the first one again [FIG. 62]. In this way fibres will grow to very considerable lengths, and the question is at once raised: what makes the nerve fibre stop growing when it reaches a suitable endorgan? It is clear that the endorgan must in some way influence the growth of the nerve fibre, probably by some chemical mechanism.

The process of peripheral nerve regeneration has been treated at some length since it is a striking illustration of growth of a cell following an injury which does not destroy the nucleus. It might be thought that this would be a much simpler affair than growth which involves cell replacement and the organization of several different tissues, but even this brief account must have shown that there are many features of nerve regeneration which are still unsolved, and plenty of scope for future work.

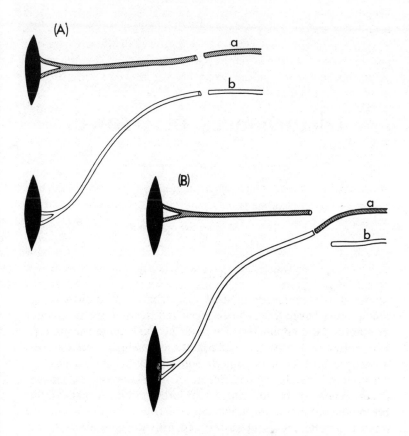

FIG. 62. Regeneration of nerve fibres.

(A) Two adjacent motor nerves of similar size are cut. (a) supplies a muscle close to the nervous system: (b) supplies one twice as far away.

(B) The central stump of (a) is sewn to the peripheral stump of (b). The fibres in (a) regenerate along the pathway provided for them by (b) and establish a functional connexion with the muscle formerly supplied by (b), having grown to approximately twice their original length. If the central stump of (b) is sewn to the peripheral stump of (a) the regenerating fibres in (b) will establish a functional connexion with the muscle formerly supplied by (a) and will then stop growing, though they have travelled only half their original length.

9

Disturbances of Growth

*'Deformed, unfinished, sent before
my time
Into this breathing world scarce
half made up'*

SHAKESPEARE

GENERAL DISTURBANCES

Some of the causes of general disturbances of growth of the whole body have already been discussed, such as the production of stunting and anomalies of growth by inadequate nutrition [p. 149]. More than fifty different conditions in which short stature is a feature have been described. Some of these are rarities, but others, for example chronic disease of the kidneys or deficiencies in the capacity of the gut to absorb certain foodstuffs, are reasonably common and important.

Two genetic causes of disordered growth have already been mentioned: Turner's and Klinefelter's syndromes [p. 147] both result from anomalies affecting the sex chromosomes. A patient with Turner's syndrome may suffer from a whole series of local growth disturbances, some of which, such as pigmented birth marks and webbing of the neck, may be of little functional importance, whereas others, such as abnormalities of the heart or of the kidneys, may threaten life. Less common than either Turner's or Klinefelter's syndromes is the condition of achondroplasia, in which the growth in length of the long bones of the limbs is markedly impaired, although skeletal maturation is normal. The skull and trunk develop normally, and so the muscles; the resulting distortion of normal body proportions is striking [FIG. 63]. The basic defect in achondroplasia is inherited as a

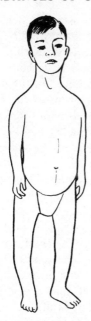

FIG. 63. Achondroplasia. Note the alterations in the bodily proportions. The head and trunk are of normal size.

Mendelian dominant. Achondroplasics have perfectly normal intelligence and were often employed as court jesters when such an occupation was fashionable.

The condition of mongolism results from a triplication of one of the autosomes, so that the mongol has a set of chromosomes totalling 47 instead of 46. The head is small, and the eyes are slanting and slitlike (hence the name); the tongue is large, and there are often congenital defects of the circulatory system. The brain fails to develop properly, and the mongol usually remains at a mental and neural age of about 5–6 years. There are sometimes other defects, which may include stunting of the whole body; the skeletal maturity is normal, but the teeth tend to erupt later than usual, and such children often die young. If they survive they become idiots who, nevertheless, are usually pleasant and tractable people.

A much more unusual condition is cleidocranial dysostosis, which affects the bones which are ossified wholly or partly in membrane. It may be considered here although very little is known about its causation; it may not be a genetic problem at all. The middle portion of the clavicle, the bones of the vault of the skull, and the mandible may be either partly or wholly deficient. When the clavicle fails to develop, the shoulders can be brought together in the midline in front of the body. If the vault of the skull remains soft and membranous, the underlying brain may be inadequately protected. Finally, there may be disturbances in the eruption of the teeth, due to the defective ossification of the mandible.

Certain genetic abnormalities affect growth less directly. For example, in phenylketonuria, a recessive defect in which the enzyme which converts phenylalanine to tyrosine in the liver is missing, phenylalanine accumulates in the body and is converted into abnormal metabolites. The child may appear normal at birth, but within a year or so there are signs of retarded mental development, and if the condition is not recognized there is ultimately a severe mental defect, associated with defective development of the brain and skull.

Again, the condition of haemophilia, which has recessive sex-linked inheritance, may cause distorted growth because of the frequency with which haemorrhages occur in relation to joints and epiphyseal plates.

The hormonal causes of growth disturbances are rather better understood. The most obvious possibility concerns the pituitary gland, which may produce either too much or too little somatotrophin [p. 144]. It appears that this may occur without the other secretions of the pituitary being affected. If too much is circulating during the period of active growth, the patient will develop into a pituitary giant. Up to the age when growth would normally cease, the proportions of the body may be relatively normal, though the growth of the limbs is relatively greater than that of the trunk. Such giants are commonly of less than usual intelligence, and many die in or before early adult life. If a pituitary giant survives beyond this time, the action of the hormone leads to continued growth of parts of the body which are still capable of growing, in particular the hands, the feet, and the jaws. The bones are thickened and clumsy; the skin and subcutaneous tissue become thick and coarse. The condition is known as acromegaly, and can occur in hitherto normal adults if a pituitary tumour develops after growth has stopped; it is apparently associated with a continuous rather than an intermittent secretion of somatotrophin. The drug bromocriptin, which in normal people raises the level of circulating somatotrophin, has the reverse effect in acromegaly, and has been used in the treatment of the condition with some success. Acromegaly is a dangerous disease with a high mortality, particularly from coronary artery occlusion.

Giantism, as might be expected, involves internal organs as well as bones, skin and subcutaneous tissue, and there is a general increase in the size of such organs as the heart, lungs, stomach, etc. The gonads, in contrast, are small. Initially the muscles are well developed, but later they become weak and easily fatigued; wasting may occur. This may result from damage to the peripheral innervation or to the muscle fibres themselves. The tongue enlarges more than the mouth, and speech becomes difficult.

Pituitary dwarfism, due to a deficiency in the amount of somatotrophin, may be difficult to distinguish from dwarfism due to other causes, such as malnutrition, congenital heart disease, rickets, etc. The skeletal proportions are usually fairly normal, and the patient simply does not grow. The facial hair commonly fails to appear, and the face may retain its childish appearance, though the skin becomes thin and often wrinkled. There is usually normal mental development, but sexual development is frequently retarded, and often the gonads do not respond at the time of puberty. If so, the epiphyseal plates of the long bones may remain open long after growth would normally have ceased, and the possibility of inducing growth in such an individual by the administration of somatotrophin is a real one, though the hormone is so difficult to obtain (only human somatotrophin is effective) that clinical experience in its use has so far been restricted.

However, a scheme for collecting human pituitary glands *post mortem* is in operation in several countries. Treatment with somatotrophin is only of use in cases where the normal production of hormone is depressed, and has no effect on dwarfism due to other causes. In other words, it is a 'replacement' therapy, and in patients with pituitary deficiency the results may be striking, growth rates of between 8 and 12 cm per year being achieved (FIG. 64). Administration of somatotrophin to an appropriate

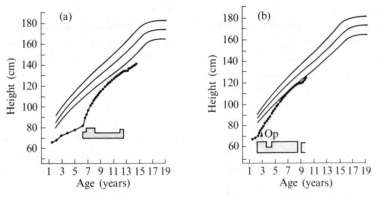

FIG. 64. Growth hormone deficiency. The patient's height is plotted on a cross-sectional centile chart, and in each figure the duration of treatment is shown by the hatched area. (A) Late treatment. Catch-up growth began immediately, but the patient failed to reach the 10th centile. (B) Early treatment. The slope of the curve alters in the same way, but this time soon enough to allow the patient to catch up to the 10th centile. Notice the transient slowing of growth after an operation (Op.). (From Mason, A. S. (1972) Short stature and its treatment, *British Medical Journal* i 519–522, by kind permission of the author and the publishers).

patient reduces the amount of subcutaneous fat, increases the thickness of cortical bone, and enlarges muscles by causing cell proliferation. The increase in height under treatment is more marked than the acceleration of bone maturation, but this is a delicate balance, and any preparation which advances bone maturation must be avoided during treatment. Patients may have to choose between delayed puberty and relatively normal height on the one hand and normal puberty and short stature on the other, for somatotrophin cannot operate on closed epiphyses.

The exact diagnosis of the nature of a failure to grow is thus of great importance in determining the kind of treatment required, and a diagnosis of a deficiency of somatotrophin is made by examining the level of the hormone in the blood. But because of the erratic and intermittent secretion of somatotrophin [p. 145] it is quite possible to find very low levels in normal children, depending on the time at which the sample is taken. It is therefore necessary to provoke the secretion of somatotrophin artificially, and to compare the response with that produced in normal children. Many tests can be used for this purpose: amino acids such as arginine may be injected, a protein feed may be given, or the patient may be made to undertake a standard exercise test. Bovril, glucagon, and other substances are also used, but the best stimulus appears to be the fall in blood sugar produced by an injection of insulin; this is also the most stressful of the methods, and is therefore not so often used as the others. The response to exercise constitutes a valuable preliminary screening test.

It is essential to reach the diagnosis of growth hormone deficiency as early as possible, for otherwise the catch-up growth produced by treatment may not be adequate to allow the patient to reach a satisfactory height [FIG. 64]. Diagnosis should be possible by the age of 5, at which time the child will have a height at least three standard deviations less than the 50th centile.

While a deficiency or an excess of somatotrophin affects the linear growth of the skeleton, a deficiency or an excess of thyroid secretion is mainly felt in relation to the maturation of bone. Disorders of the thyroid gland produce either too much secretion (hyperthyroidism) or too little (hypothyroidism). The former is relatively rare in children, but if it does occur, there is usually an acceleration of linear growth and an advance in skeletal maturation; the eruption of the teeth, on the other hand, may be slightly retarded. A hyperthyroid child who has reached the 80th centile for height may be only at the 25th centile for weight. A severe degree of hypothyroidism, such as happens when the thyroid gland fails altogether to develop, results in the production of a cretin. A cretin is a dwarf with a characteristic facial appearance—the nose is broad, the lips are thick, and the skin is thick and coarse. His skeleton to some extent retains the newborn proportions; the legs are short, and this may lead to a mistaken diagnosis of achondroplasia. The teeth erupt late, and skeletal development is also held back. The retardation of skeletal growth and maturation can be completely corrected by the administration of thyroid hormone by mouth to the child, and it is essential that this treatment should be begun as soon as possible after birth, since otherwise there may be a permanent impairment of brain function. The intelligence of untreated cretins is low, and there may also be a defect in the development of complicated movements and of speech. If nothing is done, these disabilities will persist permanently. A mild degree of hypothyroidism in a child may be difficult to diagnose, and has to be considered in all unexplained cases of failure to grow.

Thyroid deficiency in the adult leads to the condition known as myxoedema. The patient is mentally and physically sluggish, and complains of feeling cold; these symptoms are due to the general depression of metabolism, which also leads to an increase in weight. Disturbances of growth are evidenced by a tendency for hair to fall out and not be replaced, and a thickening of the subcutaneous tissue in the face; there is a characteristic loss of the hair of the eyebrows which assists in the diagnosis.

The cortex of the suprarenal glands may be overactive in childhood; in the condition known as Cushing's syndrome it produces an excess of corticosteroids, which leads to stunting as a result of interference with the secretion of somatotrophin, the deposition of excess fat, the growth of facial hair in the female, and a curious but typical 'moon face' which is also seen during the administration of large doses of cortisone to patients as part of the treatment of other diseases. On the other hand, should the suprarenal cortex produce an excess of androgens [p. 146], the result is premature sexual development. In boys, the penis, scrotum and prostate enlarge prematurely, but the testicles may remain small and undeveloped; the secondary sexual hair appears and the voice deepens. Excess or abnormal androgen production may occur during intra-uterine life, and the diagnosis of congenital suprarenal enlargement (adrenogenital syndrome) may sometimes be made at birth. In girls, the condition produces virilization; the clitoris enlarges, and there is premature development of body hair, which may follow a male pattern; the voice deepens. Menstruation and breast development do not take place.

Patients with precocious puberty of this kind are often unusually tall for their age in early childhood, and may worry about ultimate giantism. But growth stops much earlier than in normal children, and the real problem is often that of dwarfism.

Disorders of the secretion of the interstitial cells of the testicle or the ovary produce a large number of indeterminate conditions which may pose great difficulties in diagnosis. If boys are castrated before puberty, the normal accompaniments of puberty do not occur unless androgens are administered. The penis and scrotum remain small, and facial hair does not appear, though there may be some pubic and axillary hair, probably as a result of suprarenal activity. The voice does not 'break', and the patient becomes tall because of the delay in fusion of the epiphyses [p. 29]. Such patients may be either thin or fat, but usually have weak muscles [p. 147]: the prostate and other sexual organs remain infantile, and the thymus may persist. Typically the lower limbs are unusually long and the trunk relatively short.

If, on the other hand, there is excess testicular secretion, skeletal maturation is accelerated, and closure of the epiphyses occurs early, resulting in small size, the legs in particular being short. Similar phenomena occur in disorders of ovarian secretion.

Diabetes mellitus is a disease in which there is a loss of protein to the body, since it is broken down and excreted in the urine. It follows that untreated diabetes in a child leads to a stunting of growth. This can be corrected by the administration of insulin, which has the effect of increasing protein synthesis.

In the adult, a deficiency of parathormone, such as may be produced by removal of one or more parathyroid glands, leads to a deficiency of calcium in the blood and an increase in the amount in the bones. On the other hand, an excess secretion of parathormone from a tumour of the parathyroids causes calcium to be withdrawn from the bones and renders them unduly fragile and easily broken, for their strength depends largely on their mineral content. Excesses or deficiencies in the secretion of calcitonin in adults are not so well documented, but would be expected to be the reverse of those produced by parathormone. The effects on growth produced by alterations in parathormone and calcitonin secretion during childhood do not appear to be certainly established, but could be considerable.

Finally, we may mention the rare condition of progeria, in which the whole body is small, the sex organs remain infantile, the face resembles that of an old person, and the hair is white or absent. This is thought to be a complicated form of endocrine disorder, but little is as yet known about it. It is interesting that in progeria the dental age is retarded, while skeletal and mental age are said to be normal.

Despite the account given above, organic disease is relatively unusual as a cause of general growth disturbance. Much commoner is the condition of 'constitutional delayed growth', in which delayed bone maturation is associated with delayed adolescence. In clinical practice this may be difficult to distinguish from familial short stature, in which there is no retardation of bone maturation and no delay in puberty.

LOCAL DISTURBANCES

Local disturbances of growth can be divided into congenital and acquired. A detailed discussion of congenital deformities is outside the scope of this book, but it is possible to indicate their nature. In the first place there may be a failure of one or more organs or parts of the body to develop at all, a condition known as agenesis. As an example may be taken the failure of one or both kidneys to appear.

More common than complete failure of local growth is a partial failure. The testicle may fail to complete its descent from within the abdomen to the scrotum, the two halves of the palate may fail to fuse with each other (cleft palate), the foramen ovale between the atria of the heart may fail to become obliterated after birth, the two halves of the neural arches of the vertebrae may fail to unite (spina bifida), and so on.

On the other hand, there may be excess growth. Extra digits may be produced [Fig. 65(a)], supernumerary nipples are not uncommon, and a given digit may grow out of all proportion to the others [Fig. 65(b)], so producing a localized 'giantism'. Cases are reported in which half the body was much larger than the other half ('hemigiantism').

Material which should have been absorbed and removed may persist: for example, the baby may be born with a membrane blocking the anus. This represents the persistence of a stage in the development of the region, and the membrane is normally removed before birth. Tissues or organs may be found in abnormal situations; thus, one or more parathyroid glands, which are normally found in the neck, may descend into the chest, and teeth may be found in the hard palate. In the peculiar class of tumours [p. 201] known as teratomas, many different kinds of tissue may be found in strange situations, thus, teeth may be found in a tumour of the testicle, or stomach tissue and hair may be found in a tumour of the abdominal wall.

Finally, certain abnormalities arise through a process of fusion or splitting. Fingers may be fused together (syndactyly), or a single horseshoe kidney may be formed: conversely, the ureter on

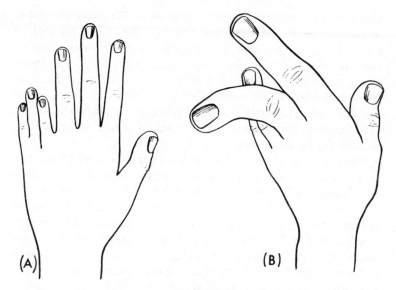

FIG. 65. Congenital deformities

(a) Supernumerary fingers
(b) Hypertrophied fingers
(Redrawn from Gould, G. M., and Pyle, W. L. (1896) *Anomalies and Curiosities of Medicine*.)

one or both sides may be double in the whole or part of its length.

The frequency of congenital anomalies deserves particular notice. About one child in fourteen that survives the period immediately following birth has a malformation of some kind; the incidence is much higher in children who die during this period. One quarter of all deaths between the 28th week of pregnancy and the end of the first year of postnatal life in England and Wales are now attributable to congenital malformations. One child in forty is estimated to have a structural defect which demands treatment. The causation of such congenital growth anomalies is only beginning to be understood; some are genetic, some may be due to physical conditions in the uterus, and others to events taking place during pregnancy. Maternal diabetes mellitus is often associated with congenital abnormalities of various kinds, and this is thought to be caused by the overproduction of insulin by the fetal islets of Langerhans, since the administration of insulin can cause deformities in animals. Viruses such as that of German measles can produce defects in the child if the mother contracts the disease at a time when the fetus is vulnerable, for example in the second month of pregnancy. Vitamin deficiencies in the maternal diet, exposure of the fetus to X-irradiation, and conditions which cause a lack of oxygenation of the blood are all possible causes. The thalidomide disaster brought to attention very forcibly the possible importance of drugs taken by the mother during pregnancy.

Some congenital anomalies are more common than others. For example, about 20 per cent of the population have a communication between the two atria of the heart. If this is small, it may produce no functional disturbance whatever, and affords an excellent example of the fact that not every anatomical abnormality is associated with a physiological deterioration. Another relatively common condition (2 per cent of the population) is a persistence of the ileal (Meckel's) diverticulum; this is usually only discovered *post mortem*, though it may become inflamed during life and mimic acute appendicitis.

The relative clinical importance of congenital abnormalities thus varies enormously. Some are automatically fatal—for example, a failure of the brain to develop (anencephaly), or a complete absence of kidney tissue. Others, such as imperforate anus, cause death if not urgently treated, and still others, though not necessarily fatal, are dangerous to life or health. Thus, a failure of the mandible to develop properly (micrognathia) interferes with normal nutrition because of the difficulty in sucking, and some congenital deformities of the cardiovascular system produce such an interference with oxygenation of the blood that early death may result.

In contrast, many congenital defects can be treated at leisure. Birthmarks, fusion of fingers, and the like come into this category. Indeed, a large proportion of vascular birthmarks regress spontaneously and disappear as the child gets older; there is no known reason for this.

Finally, some defects which are untreatable are important and others are not. For example, agenesis of one kidney can become a vital factor if the other becomes diseased, but a condition such as the existence of bilateral brachiocephalic arteries is seldom likely to worry either the patient or his medical adviser.

The system most commonly involved in congenital defects is the locomotor system, but many of the conditions which affect it are trivial; even absence of such a large muscle as the pectoralis major may produce no disability apart from the cosmetic problem. The system least affected appears to be the respiratory system.

There are differences in the sex incidence of certain abnormalities. Thus, congenital dislocation of the hip is seven times as common in females, and a congenital enlargement of the pyloric sphincter (pyloric stenosis) is about four times commoner in males.

Many congenital defects are obvious at the time of birth and others become detectable within a few days or weeks. Some, however, do not manifest themselves for a few months or even years. Among these are certain urinary tract anomalies, which make their presence known through repeated infection of obstructed urine. In the alimentary system errors in the rotation of the gut may lead to vomiting, or a deficiency in the intrinsic innervation of the colon may result in severe constipation and an enormous dilatation of the gut (Hirschsprung's disease). In this group is spina bifida occulta, in which, following a failure of the fusion of the neural arches of one or more lumbar vertebrae, the dura mater is connected to the skin by a fibrous cord. In such cases the differential growth of the vertebral column and the spinal cord may result in the nerve roots and the cord itself being pulled upon, so that in later childhood or adolescence signs of spastic paralysis and loss of bladder control may appear.

Before proceeding to consider acquired local disturbances in growth it is necessary to discuss some definitions. The first of these is the term hyperplasia, which is applied to a local increase in the number of cells in an organ or tissue, such as occurs in the gonads at puberty. This process is reversible; we have already seen [p. 14] that the hyperplasia of the blood-forming cells in the bone marrow following severe blood loss regresses again once the blood loss has been made up.

The number of cells in a unit volume of tissue can theoretically be increased in three ways. Prolongation of the life span of the individual cells, acceleration of differentiation, and an increase in the rate of cell division will all have this effect; the latter is the usual way in which hyperplasia is produced.

It is perhaps difficult to realize the distinction between hyperplasia and the normal growth of cells in an adult tissue which was considered in CHAPTER 8. The difference may be appreciated by considering once again the situation in the skin. The thickness of the epidermis remains more or less constant, because there is an equilibrium between the rate of division of the stem cells and the rate of removal of the superficial cells [p. 13]. Now let us suppose that there is an increased shedding of dead cells from the surface because of some disease process or mechanical attrition. This, by a complex chemical mechanism, of which we know very little, influences the division rate in the germinative layer so that a new equilibrium is reached. More cells are now to be found in the germinative layer than would be there in normal circumstances, and this is what is meant by hyperplasia. The converse of hyperplasia, in which fewer than the normal numbers of cells are present, is called hypoplasia.

The next term which requires definition is hypertrophy. This means an enlargement of cells without an increase in their number. Increase in the size of its component cells naturally results in an increase in the size of an organ or tissue, which is then said to be hypertrophied. Hypertrophy may result from an increased functional demand. Thus, if there is a gradually increasing obstruction to the outlet of a hollow organ, the muscle which expels the contents of the organ will hypertrophy. A narrowed valve in the heart may thus result in a thickening of the cardiac muscle fibres, perhaps to double their original diameter, and this is reflected in a great increase in the thickness of the wall of the chamber of the heart primarily concerned. Similarly the wall of the bladder thickens greatly when the urethral outlet is progressively blocked by an enlarged prostate gland. Somatic muscle can be made to hypertrophy by exercising it [p. 36], and what happens is that the smaller fibres in the muscle are brought up to the normal average size; those which are already large do not appear to enlarge further.

Again, when a blood vessel is blocked, the small collateral branches which unite with similar branches of the neighbouring blood vessels will enlarge to carry the blood past the obstruction [FIG. 66]. This enlargement is progressive, and takes place over several days; the stimulus which causes it is thought to be the increase of pressure inside the small vessels, and in this case the

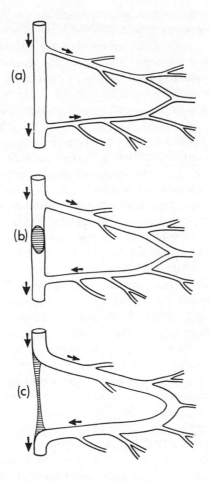

FIG. 66. Development of collateral circulation.

(a) Normal anatomy. Little blood flows along the anastomotic connexion between the two collateral branches of the artery.

(b) Clot forms in the main artery between the two branches. Pressure beyond the site of blockage falls, and blood now passes through the anastomosis and back to the main trunk.

(c) Revised anatomy. The main artery is permanently obliterated by fibrous tissue and shrinks to a fibrous cord. The increased pressure has caused the anastomotic channel to dilate; eventually its walls thicken by hypertrophy and hyperplasia and it becomes a vessel indistinguishable from the original main artery. The blood is now supplied to all parts of the territory formerly supplied by the main artery just as efficiently as before.

hypertrophy of the wall is accompanied by a hyperplasia of the cells composing it. A similar combination of hypertrophy and hyperplasia is seen in the uterus and the breasts during pregnancy, when it is attributable to the effects of various hormones.

Finally, a condition called compensatory hypertrophy develops when an organ such as the kidney or the ovary is removed. The remaining kidney or ovary enlarges—the ovary possibly even to the size of the combined mass of the ovaries before operation, and the kidney perhaps to 80 per cent of the original combined mass. If the functional units of an organ undergoing compensatory hypertrophy are single cells, the enlargement of the whole organ may be due to hyperplasia of the active cell population, as in the suprarenal cortex. If the organ contains more complicated functional units, such as the nephrons of the kidney, these cannot apparently be multiplied beyond the original normal adult complement, and a certain amount of cellular hyperplasia is supplemented by cellular enlargement, both processes resulting in a more or (usually) less complete restoration of the number of functional units and some increase in their size. In this way much of the functional deficiency is made good—so much so that removal of 80 per cent of the total kidney substance is not necessarily fatal. In the ovary most of the compensatory hypertrophy is the result of follicular growth and maturation, and in the testis, which does not enlarge to anything like such an extent after a unilateral removal, the increase in size is mostly due to the growth of the interstitial tissue.

The converse of hypertrophy is atrophy. This term is used to imply either a loss of cells or a diminution in their number, or both, and is also applicable to the organ or tissue in which this happens. Atrophy may follow disuse; if a limb is immobilized in plaster of Paris and the muscles are not kept exercised by physiotherapy, they will be found to have shrunk when the plaster is removed. Malnutrition produces atrophy, as can be seen in famine victims. Lack of sufficient blood supply, as in the heart wall when the coronary vessels are becoming narrowed [p. 227], can cause atrophy. Denervation of muscles results in their atrophy, for reasons which are not completely understood; it is probably due to something more complicated than simple disuse, and trophic factors may be concerned [p. 143]. Finally, endocrine disorders may produce local atrophies. For example, in

Simmonds' disease, in which the anterior lobe of the pituitary gland is destroyed, there is a reduction in the size of the heart, with atrophy of the gonads, the suprarenal cortex, and the breasts.

These facts raise once again the question of the control of growth in a normal tissue or organ. In any biological material the balance between anabolism and catabolism is the fundamental factor which determines the mass of the material at any given time, and it has already been postulated that the feedback mechanism which controls this balance is a chemical product of the active cells [p. 14]. But clearly the response of a tissue to increased or decreased functional demand, its reaction to denervation, and the results of endocrine disturbances, indicate that there may be other chemical mechanisms of which we know virtually nothing. Probably the range within which the balance operates is determined genetically, and within this range the chemical mechanisms constitute the 'fine adjustment'.

Another term which requires discussion is metaplasia. This means a replacement of one tissue by another; the new type of tissue is thought to come from cells that have never completely differentiated, and so have retained many of their original wide potentialities. This implies that adult connective tissue (for example) still contains a few of the original stem cells. A common type of metaplasia is seen when the ciliated respiratory epithelium in the air sinuses or the bronchi is replaced by squamous epithelium as a result of prolonged surface irritation. Another common example is the appearance of cartilage in tendons subjected to localized friction. Metaplasia is a sign which may be a precursor of neoplasia.

Neoplasia differs from normal cell division and from hyperplasia in that it is progressive and continues irrespective of external stimuli. If once again we use the skin as an example, what happens in neoplasia is that cell division in the germinative layer escapes from control, and no longer bears any relation to the rate of loss of material from the surface. The result is a local uncontrolled hyperplasia, resulting in an increase in size of the epidermis, which is called a tumour or neoplasm. The essential difference between growth in a tumour and growth in normal skin is that no limit is set to tumour growth other than the death of the host.

Neoplastic cells are derived from ordinary tissue cells which, in a manner not yet understood, undergo an abnormal type of irreversible change. When neoplastic cells divide, they transmit to their descendants the same abnormal quality they have acquired, so that the family of cells they produce loses its normal growth-controlling mechanism. Nevertheless, some of the characteristics of the normal tissue cells may persist in some of the members of the family. In any tumour there are many stem cells, and mitotic figures are common, but some of the daughter cells become differentiated to a greater or lesser extent, and may carry out functions typical of their heritage. For example, many of the cells of a tumour derived from the thyroid gland may continue to secrete thyroid hormones.

The proportion of stem cells and relatively differentiated cells varies according to the particular tumour and the rate at which it is growing; fast-growing tumours have naturally many stem cells, but slow-growing ones may resemble normal tissue quite closely on superficial microscopial examination. One of the reasons for the excessive growth of tumours may be that the stem cells in the tumour do not mature and differentiate as quickly as they normally should, and therefore remain capable of dividing for a greater proportion of their lifespan. Tumour cells do not divide any more rapidly than normal tissue cells, and their mitotic cycle may even be slower.

Some tumours are classified as benign; that is, they do not invade the surrounding tissues, and produce their effects merely by their size and the consequent pressure on adjacent structures, or by the functional results of the activity of their differentiated cells. Other tumours become malignant, and their cells proceed to infiltrate and destroy surrounding tissues; they are also capable of forming daughter tumours (metastases) in distant sites. This happens if a portion of the tumour becomes detached and carried away by blood vessels or lymphatics to settle elsewhere in the body; it is a feature of malignant tumours that such nests of cells are capable of further growth in new, and often apparently unsuitable, surroundings.

Cells in malignant tumours present certain morphological features which enable them to be recognized, and which are collectively described as anaplasia. Strictly speaking this term signifies merely a lack of differentiation, but associated with this the orientation of the structures within the cells is often disturbed, and the shape and size of the nuclei tend to vary. Mitoses occur in large numbers, and are often abnormal in appearance; the chromosome number is often irregularly altered (aneuploidy). The organization of the cells into a tissue is also abnormal, with variations in structure in different parts of the same tumour. Sometimes the growth of the tumour outstrips its blood supply, and many of the cells within the tumour can then be seen to be dead or dying. In other cases there is no apparent breakdown of blood supply, and it is suggested that in such circumstances the death of cells occurs because the mass of the tumour, is, like the mass of an ordinary tissue, controlled by a self-regulating chalone mechanism [p. 14].

The causes of local disturbances of growth are numerous and varied. It has already been said [p. 170] that local growth is stimulated by injury, and occasionally the process may be rather too enthusiastic, so that more scar tissue is formed than is needed to repair the defect. Such an overgrowth of collagenous tissue is known as keloid, and for some reason the condition appears to be commoner in Negroes and during pregnancy. The overgrown scars which sometimes follow acne are considered by some to be a type of keloid. Very little indeed is known about the genesis of keloids, and they may appear spontaneously in uninjured skin.

Local irritation of the skin, especially when intermittent, may produce a local overgrowth such as a corn, which is confined to the site of pressure. This is in one sense a kind of tumour, but there is no uncontrolled growth of the germinative layer, and the corn will regress and disappear if the irritation is removed. On the other hand, if certain types of irritation are continued too long, an uncontrollable neoplasia may be stimulated, and eventually spreads beyond the site of the stimulus (local malignancy) or to distant parts of the body (general malignancy). Removal of the original irritative cause will have no effect on the further growth of the tumour.

The first example of this sort of thing to be generally recognized (in the eighteenth century) was the cancer of the skin which at that time was common in those people who worked with tar. A great deal of research has since been done on the role of chemical irritation in cancer, and in 1928 Kennaway discovered that 1.2.5.6 dibenzanthracene, isolated from tar, possessed the ability to induce cancer when painted on the skin. Many other such carcinogens have since been identified.

As is now well known, there is an association between cigarette smoking and cancer of the lung, but so far efforts to identify a specific carcinogen in cigarette tar have not been successful. It may be that in this case the irritation is a general one rather than due to a specific chemical, for many other examples of this kind are known. Cancer of the stomach may follow a long-standing stomach ulcer; cancer of the tongue may result from smoking a clay pipe, and cancer of the skin is common in countries, such as Australia, where the skin is exposed to excessive sunlight. It does not seem to matter what the irritant is; provided it is adequate to produce a neoplastic change, the subsequent course of events is similar in all cases, though the actual tumours naturally have somewhat different characteristics, because they develop from different types of cells.

Another cause of neoplasms is infection by viruses, which may produce either benign or malignant tumours by disturbing metabolic processes within the cells. A familiar example of a virus-induced benign tumour is the ordinary cutaneous wart. Some malignant tumours of animals have been proved to be due to viruses, but whether or not this is an important cause of malignant neoplasms in man remains to be seen.

Genetic factors are also involved, and it is often the suitable 'soil' in which the neoplasm can take root which is inherited. To take a specific example, there is an inherited tendency to develop small benign tumours (polyps) of the large intestine, and these have in turn a tendency to develop malignancy in later life. There are, however, some inherited tumours which are in themselves highly malignant, such as retinoblastomas, which are genetically determined tumours of the retina.

Hormones are known to influence established neoplasms of particular organs; cancer of the prostate is susceptible to treatment by oestrogens, and cancer of the breast can be made to regress temporarily by removal of the ovaries. There is now good evidence that administration of excess oestrogens may sometimes lead to cancer of the breast, and it is probable that hormonal factors may be concerned in the production of other tumours.

Radiation can also induce malignant tumours. The classical example occurred in a factory in which radioactive material was used to paint luminous figures on watch dials. The girls doing this used to moisten their brushes with their lips, and in this way took in an amount of radioactive material sufficient to produce malignant tumours of their bones.

Lastly, we mention the wholly mysterious effects exerted by the central nervous system on certain forms of neoplasm. It is well known that warts may be 'charmed' away by hypnosis or other forms of suggestion, but the processes underlying such recoveries are quite obscure.

10

Old Age

'There are so few who can grow old with a good grace'

STEELE

NATURAL HISTORY OF SENESCENCE

We live in an ageing population. In 1977 there were 9·4 million people in Britain over the age of 65, and if present trends continue, this figure will rise to 9·7 million by 1986. The number of people over 85 is expected to increase by 50 per cent by the year 2000, and it is estimated that there will by then be 3·6 million aged 75 or over. It follows that the structural and functional changes which occur in old age are becoming progressively more important as a subject for study by the medical profession and by all those who have to deal with old people. These changes, which are collectively known as 'senescence', do not appear to any extent in animal populations under natural conditions, since individuals are killed and eaten before the more noticeable structural changes have time to develop. In primitive human communities conditions were probably not so different, and it follows that senescence is a relatively modern phenomenon, which even now is only seen in fully developed form in the relatively sheltered conditions surrounding old age in communities with a reasonable standard of living.

Senescence is a difficult word to define, since it is impossible to distinguish with any firmness between changes inherent in the mere passage of time and changes due to degenerative disease: a survey taken in the United States in 1962 estimated that approximately 80 per cent of people over 65 are afflicted with one or more chronic illnesses. Nor is it established whether the diseases of old people are the cause or the result of what we describe as senescence. It is possible that there is a genetically determined 'curriculum vitae' which causes the onset of senescence quite independently of wear and tear and of disease.

In general terms senescence can be considered as a stage on the pathway to death—a stage which, because of the disabilities it brings, itself increases the likelihood of death occurring. Nevertheless, a civilized society is to some extent cushioned against the effects of many of these disabilities. For example, for a wild carnivore the loss of its teeth means certain death, but for man edentulousness may result merely in some difficulty over eating particular foodstuffs. The dangers of other disabilities, such as prolongation of reaction times [p. 211] and motor weakness, can be mitigated by taking appropriate environmental precautions, such as avoiding crossing busy streets.

Countries can be roughly categorized according as their population is 'old', 'middle-aged', or 'young' by estimating the percentage of people aged 65 or over. Thus, France is an 'old' country, with over 10 per cent of her population in this group, and parts of Africa are 'young' in that less than 4 per cent of the population have achieved this age. FIGURE 67 shows the alterations in the life tables for the United States which have taken place this century.

An artificial classification adopted by the World Health Organization regards 'middle-age' as the period from 45 to 59. Those aged 60 to 74 are classed as 'elderly', those aged 75 and over as 'old', and those aged 90 or over as 'very old'.

Just as the best way of growing tall is to have tall parents, so the best way of living to an advanced age is to have long-lived parents, although the correlation between longevity of parents and longevity of children is not so good as that between tallness of parents and tallness of children. It is also very advantageous to be a female, for 'old' women and 'very old' women considerably outnumber old men. Thus, in 1963 more than twice as many females as males had reached the age of 80 or more.

What has been called the 'survival ability' of the fertilized ovum varies enormously; some die during early development in the uterus, others survive only for a short period after birth, and a few—those who have inherited genes particularly capable of looking after the individual during the vicissitudes of life—battle

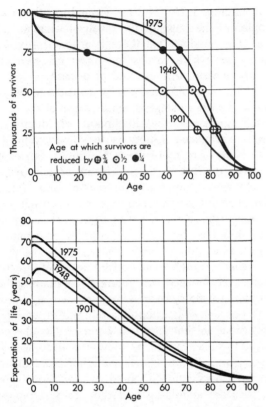

Fig. 67. Expectancy of life. The graphs show the actual expectancies for 1901 and 1948, and the projected expectancies for 1975. Note that the expectancy is considerably increased for children and young people, but not for older people; the total span of life does not appear to be increasing, but the numbers of people in the older age groups is rising steadily. (From Comfort, A. (1964) *Ageing: The Biology of Senescence*, 2nd ed., London, Routledge and Kegan Paul Ltd., by kind permission of the author and publishers.)

through to old age. Even they are eventually defeated, usually before reaching the age of 100. The maximum life span supported by a birth or baptismal certificate appears to be 142 years, achieved by a man in the high valley of Vilcabamba in Ecuador, where many people reach great ages. They attribute their longevity to a low calorie diet which is largely vegetarian and almost completely lacking in fat, to the beauty, peace, and equable climate of their valley, and to the consumption daily of 2–4 cups of unrefined rum and 40–60 cigarettes! There are tales of greater ages, and the oldest man in Azerbaijan in 1968 claimed to be 163 years old. Most such cases are vulnerable because of the absence of birth certificates or of proof of identity. Although the record seems to be held by a man, women centenarians in Britain currently outnumber male centenarians by about 6 to 1; no reason for the sex discrepancy has yet been scientifically proved, though there are many suggestions. It may be, for example, that the Y chromosome is responsible for shortening male life, as it appears to be responsible for delaying male maturation [p. 142]. If so, patients with the constitution XYY might afford useful information, but too little is yet known of their life span compared with normal males.

The expectation of life of old people is not increasing anything like so fast as the expectation of life of young people. There is little prospect that in the near future everybody will live to be over 100, and centenarians will continue to be exceptional people. What does appear likely is that more and more of the population will survive into the octogenarian range, since medical research, though it has not mastered the problems of ageing, has been able to prevent early death and to prolong the expectation of life of those in middle age.

The changes of senescence occur for the most part after the end of the child-bearing period, and it is therefore probable that resistance to them has not yet evolved very far, since natural selection could not operate directly on such a situation. It is, however, coming to be realized that the changes of old age do not appear simultaneously in all systems; just as in youth many systems have different growth curves, so in the later part of life many systems have different curves of deterioration. It may even be said that old age begins to make itself felt as soon as maturity is reached, for by then the first structural and functional changes are apparent in the nervous system. Natural selection could conceivably operate on these early symptoms and signs of old age, which render their possessors marginally less fit than their competitors.

If, then, we spend about a quarter of our lives maturing and the remaining three-quarters in growing old, it is perhaps rather surprising that the study of the changes in old age by scientific methods is among the youngest disciplines in medicine. One reason for this is that it is only quite recently that old people showing such changes occurred in sufficient numbers to constitute a major medical problem. There have as yet been no comprehensive longitudinal studies of ageing, and the difficulties in such an investigation are even more formidable than those in similar studies of growth and maturation; the existing cross-sectional data are open to the usual objections [pp. 16, 153].

Nevertheless, it can confidently be said that many of the chief disabilities of old age are due to changes in the cells and tissues of the body, prominent among which are defects in the processes of growth and of replacement of damaged cells.

CELLULAR CHANGES IN AGEING

In an attempt to find out why tissues eventually die; Alexis Carrel cultured chick fibroblasts, and found that they could be maintained and subcultured for twenty years, which was much longer than the life span of the hen from which they were derived. He concluded that ageing was not due, as had been thought, to a decline in the capacity of the stem cells to reproduce themselves [p. 12]. In Carrel's experiment the cells were fed a crude extract of chick embryonic material, and it is now thought that a few living chick cells must have been added to the culture with the extract, for if living cells are rigorously excluded from the preparation, it does not survive long.

Mouse cells and human cells have been made to grow in culture for many years (the so-called HeLa strain of cells was derived from the uterine cervix of a patient in 1952, and is still growing), but only after the cells have undergone nuclear changes which transform them into abnormal cells. For example, the HeLa strain now has 50 to 350 chromosomes per cell, and it has been found that many cells of long-lived lines will grow as malignant tumours if injected into a suitable host of the same species. Somewhere along the line of the innumerable subcultures it appears that the originally 'normal' cells develop qualities similar to those of neoplastic cells, which are easy to grow in tissue culture, and can be maintained pretty well indefinitely.

Human fibroblasts have a finite lifetime in tissue culture: a sheet of fibroblasts grown on nutritive medium will double itself from 40 to 60 times and then the cells in it will die. Cells removed from the culture after 30 such 'doublings' and put into cold storage will, if revived, 'remember' the stage they had reached, irrespective of the time interval, and will go on to divide about 20 more times. Towards the end of the lifetime of such a culture aberrations in the chromosomes begin to show up, and there is a gradual decline in the capacity for reproduction.

Such findings may not have a direct relevance to the behaviour of cells growing in their natural surroundings in the intact body, but the evidence is at least suggestive that the same sort of thing may happen. Neoplastic growth is commoner in older people than in the young, and cancer begins to obtrude itself as a cause of death. At the age of 25 a man has a $1:700$ chance of developing cancer in the next 5 years; when he is 65 the chance is $1:14$. Yet if cancer does occur in old age, the growth capacity of the cancer cells is often not as great as it is in a younger person with the same type of tumour, and consequently such neoplasms tend to be slower growing and less invasive.

For the purposes of the present discussion, we are less concerned with local neoplasia than with generalized atrophy, which is a characteristic feature of old age. We have already seen that cell division as a means of growth is most important in fetal life and early childhood. After puberty, when maturity has been reached, mitosis persists to keep the tissues and organs in a steady state as regards cell numbers. With advancing years mitosis begins to fail to keep up with attrition, and a phase of negative growth begins, in which there is a decline in the number of cells in the body. (The curve of this decline does not run parallel with the mortality curve, and the inference has been drawn that the decline in cell reproducibility cannot be the *only* cause of senescence.)

The first manifestations of the progressive loss of cells are naturally to be found in the central nervous system, where cell replacement never occurs after birth [p. 55]. They become noticeable shortly after the adolescent spurt has been completed, for it is then that the special senses start to lose their acuity and range of performance. (Hearing is said to be most acute round about the age of 10 years.) At this time also, reaction times begin to lengthen, particularly those involving multiple choices of reaction, and co-ordination starts to suffer. These effects are manifest about 20 to 25 years of age, and proceed progressively throughout life. The special senses continue to deteriorate, though the effects are at first too gradual to be noticed unless accurately tested. In later life the functional deterioration is accompanied by anatomical signs of cell depletion. Thus the olfactory bulb of the brain atrophies, and there is evidence of atrophy of the nerve cells concerned with the perception of high notes. The anatomical evidence in relation to vision is more complex, and it must be stressed that in this sense as in all the others the atrophy of nerve cells is only one factor in the decline of performance. The first sign is usually a reduction in the capacity of the eye to accommodate about the age of 16 to 17 years, and this is probably nothing to do with loss of central control. Speech becomes slower and less assured in old age, probably because of atrophy of the central nervous mechanisms involved.

Memory, and the ability to learn, are already much less efficient in early adult life than they were in childhood, and in general it may be said that the trade union principle of 'last in, first out' applies in at least a limited way to the nervous system, since functional defects are first apparent in relation to the cell assemblies which control the most complicated and most recently acquired functions. It is now a commonplace that athletes become 'too old' for many competitive sports by the age of 35, particularly those requiring tremendous co-ordination, such as ice-hockey, tennis, and so on. The effects of age on professionally acquired skills—apart from the decrease in physical strength—is a tendency towards slower, though not necessarily less accurate, working. The speed of problem solving decreases from early adult life onwards.

In later life the progressive loss of nerve cells leads to some loss of control over the upright posture—a control so laboriously acquired over many of the early years of life. Thermal regulation by the nervous system may fail, and there is danger from exposure to cold: the autonomic system loses some of its control over smooth muscle, and incontinence may result. The autonomic deterioration is also responsible for the fact that the systolic blood pressure may fall by up to 20 mm Hg or more in about a quarter of otherwise fit elderly people on standing up. This condition of postural hypotension [see p. 139] is one cause of the 'drop attacks' (sudden loss of consciousness and falling) which are common in old people.

These symptoms and signs are accompanied by a diffuse atrophy of the brain, particularly in such regions as the frontal lobes. The fissures are deeper and wider than in youth, the weight of the brain is less, perhaps as much as 10 per cent, and there is a relative increase in the neuroglial tissue. Some of the cells which remain show signs of abnormality. For example, their nuclei may be irregular, and may stain more darkly than normal. Many cells appear to shrink, and though this is difficult to establish with certainly there is reasonable evidence that shrinkage occurs in the Purkinje cells of the cerebellum and the pyramidal cells of the motor cortex. In senile cells the Nissl substance is said to be depleted, and increasing amounts of a yellow-brown pigment known as lipofuscin may be deposited in both nerve cells and glial cells; the accumulation of pigment may push the nucleus to one side. It is thought that these 'lipofuscin bodies' may be the end-stage of the life-cycle of the cellular lyzosomes: in cells such as the nerve cells these have no chance to get rid of their contents by voiding them from the cell, and they may ultimately poison themselves by the mere accumulation of waste material. Others believe that the lipofuscin is derived from the degeneration of mitochondria.

The organization of the dendrites may become more complex, and short new dendrites may be formed, scattered over many parts of the cell body; at the same time structures known as argyrophilic plaques and neurofibrillary tangles may appear. It should perhaps be added that many of the microscopic changes described in the nervous system in relation to ageing are difficult to evaluate, since there has been little effort to quantify the results obtained, to exclude disease, and to avoid artefacts due to post-mortem changes or differences in experimental technique.

It has been calculated that the brain loses something like 10 000 nerve cells every day, and that by the age of 65 to 70 about 20 per cent of the total number of neurones present at birth have been lost. However, the evidence for this estimate is unsatisfactory and conflicting, and recent work has shown that the nuclei of the brain stem retain their complement of cells into advanced old age. Again, the numbers of cells in the temporal cortex and the numbers of Purkinje cells in the cerebellum appear to remain fairly steady until the age of 60–65 years, and only then do they begin to diminish. A decrease with age in the number of fibres in the dorsal roots of the spinal nerves has been reported after the third decade, and in the spinal cord and posterior root ganglia atrophy and pigmentation are progressive, so that after the seventh or eighth decade few cells in the ganglia are unaffected.

The peripheral nervous system also suffers, though not to the same extent. The maximum speed of nerve conduction in large trunks such as the median and musculocutaneous nerves decreases with age, but the speed of simple reflexes does not seem to alter much in old age, nor, so far as the evidence goes, does skin sensibility. But the number of Meissner corpuscles in the skin decreases with age, and the survivors increase in length, becoming coiled and lobulated. In the same way, the number of taste buds in the tongue drops from about 250 in each circumvallate papilla to a figure less than 100. Partly as a result of this, partly because of central losses, and partly because of a reduction in the amount of saliva secreted, the sense of taste is impaired in old people; in those over 80 the electrical threshold for taste is about five times higher than it is in those in their twenties. Smell is also said to be impaired in the elderly.

Although the nervous system is the first, and perhaps the most striking victim, a similar kind of defect becomes apparent in all the other systems with the passage of time. The lymphatic system is one of the first to show changes, for many of its components progressively atrophy and disappear in early adult life, especially the part of it associated with the respiratory and alimentary systems and sometimes called the epitheliolymphoid system. The thymus gland has usually largely been replaced by fibrous tissue even before complete maturity has been attained, but complete loss of thymic tissue is unusual, even in old age. The germinal centres of the lymph nodes begin to retrogress at puberty, and the structure of the nodes begins to alter; the cortex becomes thinner, and the distinction between cortex and medulla less evident. In extreme old age the structure of some nodes becomes more or less uniform throughout. The lymphoid component of the spleen begins to decrease about puberty, and continues to do so more gradually until old age, when the total weight of the organ begins to fall more rapidly than the body weight.

The other obvious tissue in which to look for age changes is somatic muscle, for, like nerve cells, muscle cells do not add to their numbers after birth, and each muscle cell thus lives (potentially, at least) as long as the body lives. However, unequivocal evidence of senile, as opposed to pathological, changes in human striated muscle is hard to find: much of the work has been done on the extraocular muscles, which in most people are subjected to a fairly standard amount of activity, and which tend to be relatively free from gross pathology. The changes observed include alterations in the arrangement, pattern, and integrity of the myofibrils, deposition of lipofuscin pigment granules, and an increase in the collagen and elastin content of the muscles. Atrophy of individual fibres is not a feature, and indeed it appears that the average size of the fibres may actually increase. This may possibly represent an attempt on the part of surviving fibres to compensate for the loss of fibres which is presumed (though not proved) to occur. Somatic muscle in other situations seems to preserve its structure remarkably well until extreme old age. It has been suggested that the primary cause of the senile changes in muscle is a partial failure of the trophic mechanisms of the nerves which supply it [p. 143]. However, motor nerve cells also die and their motor units atrophy, the blood supply to muscles decreases, and the production of testosterone, which is responsible for the increase in muscle at puberty [p. 147], declines. The net result of these and possibly other factors is that the total weight of muscle in the body falls by about 30 per cent between the ages of 30 and 90.

In the same way other tissues and systems lose cells because the growth processes come to be inadequate to effect normal running repairs. This shows in several ways. In an old person there is a loss of subcutaneous fat, and the epidermis becomes thinner. The extent to which this occurs is difficult to establish; figures for the epidermal thickness in the antecubital fossa show an average loss of some 6 micrometres between the ages of about 30 and 85, but some of this difference may be due to a loss of elasticity; in a young person the epidermis is 'bunched up' by the pull of the elastic fibres in the dermis, and so appears to be thicker than in an old person, in whom the elasticity of the dermis is reduced [p. 222]. If the epidermis is separated from the dermis before measurement the difference in thickness with age is considerably less, but this procedure may in itself introduce other complications.

What is certain is that the number of dermal papillae decreases continuously after middle age, and they also become flatter, with a loss of the rate pegs of the epidermis. As early as the age of 30 or so the replacement of terminal hairs by vellus on the vertex and in the temporal regions may become noticeable, and by 50 about 60 per cent of males have some degree of baldness on the vertex; the number of hair follicles becomes progressively reduced with advancing age. However, there is an unexplained tendency towards increased growth of coarse hair in the nostrils and the external ears. At the same time faulty growth processes in the remaining hairs lead to their becoming grey or white. The rate of growth of finger-nails [p. 164] declines steadily through later life, and the nails become brittle and crack easily.

The reproductive system deteriorates in middle age. The ovaries, scarred by oocytes escaping over the years, begin to atrophy. They may contain cysts and show various degenerative changes (p. 224). Follicles cease to develop, and viable ova are not usually produced after the age of about 50 years. Because no new oocytes are formed after birth [p. 94], the oocytes which are shed in the later stages of reproductive life may be as much as 40 or more years old. This is perhaps significant in relation to the fact that the incidence of mongolism [p. 187] increases with increasing maternal age. Much of the ovarian stroma becomes converted into fibrous tissue, so that in old age the ovary may be represented by a mere strip of fibrosed material.

The menopause, which marks the cessation of the ovarian and uterine cycles of growth, death, and repair [p. 166], occurs at a very variable time in different women, but it is rare for menstruation to persist in a normal cycle much beyond the age of 50. However, just as the menarche is occurring earlier, so the menopause appears to be occurring later in life; a century ago it seems to have been four or five years earlier than it is now. The diminution in the amount of oestrogens secreted by the ovaries has its effect on the whole of the rest of the reproductive apparatus. The uterus shrinks, its mucosa becomes atrophied, and the glandular tissue largely disappears. In old age it may be converted into a small fibrous body quite unlike the mature organ. The vaginal epithelium thins, and there is some drying and keratinization of the vagina and vulva; the reaction of the vagina changes from acid to alkaline. Breast tissue atrophies after the menopause, probably because of the lack of ovarian stimulation, and in old age there may be little glandular tissue left, while there is also a marked diminution in the fat surrounding it [FIG. 37]. As a consequence the skin over the breast sags in wrinkles.

In the male there is no abrupt cessation of the production of viable spermatozoa, which still continues, in at least 50 per cent of subjects, to the age of 70 years or over. There are many accepted cases of fertility in males of over 80 or even 90 years. But changes in the testicle can be detected much earlier, and consist of a reduction in size, a thickening of the basement membrane of the seminiferous tubules, a reduction in the number of interstitial cells, and an increasing fibrosis in the intertubular areas. Lipofuscin pigment tends to deposit in the interstitial cells and in the lining of the ductuli efferentes, and there is a gradual but considerable reduction in the amount of hormone secretion in the sixth decade. As in the female, this is probably responsible for many of the changes in the rest of the male sex apparatus. Thus, in the prostate gland there is an atrophy of the smooth muscle component relative to the connective tissue, the glandular component undergoes patchy atrophic change, and pigment is deposited in varying degrees as spheroidal granules in the smooth muscle cells. In the ducts concretions called corpora amylacea make their appearance. These changes become more marked with increasing age, and by the age of 70 they may be universal throughout the gland. Curiously, the prostate is an exception to the general rule that the weight of organs decreases in the elderly. It *may* become smaller round about the age of 55 to 60 years, but in many men this is the time when it starts to hypertrophy rather than to atrophy. Ten years later the enlargement of the median lobe of the prostate may constitute an increasing obstruction to the passage of urine out of the bladder. It is said that, after the age of 70, two out of three men have some degree of urinary obstruction from this cause. By this time also there may be a decrease in the size of the seminal vesicles and in the thickness of their lining epithelium, in which yellowish pigment accumulates. There is, however, no consistent relationship between the weight of the prostate and that of the seminal vesicles.

Changes in the endocrine system have been reported in old age, but they are often difficult to detect, and their significance is obscure. For example, in the suprarenal medulla there is an increase in the amount of connective tissue, which forms a coarse meshwork round the parenchymal cells; there is often some dark pigment resembling melanin. In the cortex there is an increase in the number of cells containing yellow or brown pigment, and in very old people there may be conspicuous clumps of such cells. Most other endocrine glands have been said to exhibit an increasing fibrosis with age.

The teeth deteriorate through disease and tend to fall out; the gums recede, and the bone of the jaws is absorbed, while the maxillary sinus [p. 77] enlarges. The absorption of the alveolar margin leads to the angle of the mandible [p. 77] becoming more obtuse, reaching a figure of about 140 degrees. There is a loss of fat in the orbits, so that the eyes become sunken, and the weakness of the eye muscles affects the processes of focusing and convergence. After the age of about 60 the progressive loss of kidney tissue leads to an overt impairment of excretory function, and there is a diminution in the basal metabolic rate. Other internal organs, such as the pancreas, liver, heart, and spleen, also tend to become smaller [FIG. 68]. Anaemia is common in the elderly, and this may be partly due to inefficient replacement of worn out red blood cells. It is difficult, though, to dissociate the anaemia from the inadequate diet of many old people. Old age pensioners who look after themselves exist largely on tea and toast, and such a diet is grossly deficient in the iron necessary for forming haemoglobin.

The generalized atrophy of tissues and systems is particularly evident in skeletal muscle [p. 215], and this usually leads to a marked loss of weight and contributes to the typical 'wasted' appearance of the very old. On the other hand, the deposition of fat in middle age tends to mask this, and some old people still eat enormous quantities of food, with the result that their body weight may actually increase. The composition of such a person's body is necessarily very different from that of the young adult [p. 38].

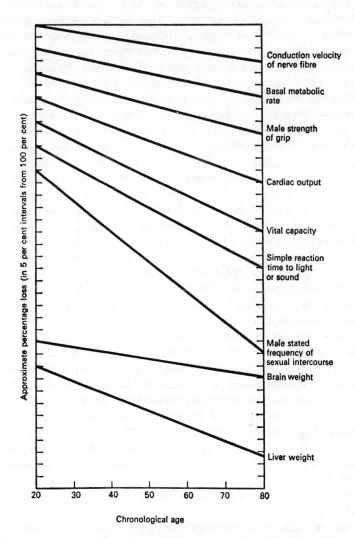

Fig. 68. Simplified diagram showing deterioration of various biological measurements, expressed as percentage loss of function from the age of 20. (From Bromley, D. B. (1966), *The Psychology of Human Ageing*, Penguin Books, Harmondsworth, by kind permission of the author and publishers.)

The blood urea tends to rise, as do the blood cholesterol, sugar, creatinine, and uric acid. The white cells diminish in number because of underproduction of lymphocytes, and the blood sedimentation rate rises. Most of these changes are manifestations of changes in the functioning of cells and systems.

It might be expected that the diminished capacity of cells to undertake division would be reflected in a lessening of the ability to repair damaged tissues, and this is borne out by observations on the healing of fractures. In the elderly, the bones are brittle [p. 224], and it is thus common for them to break. It is also unfortunately common for them not to heal properly, and many old people with a fractured femur may never succeed in bearing weight on it again. The exact reason for failure to achieve good bony union is not, however, certainly known. It may in part be due to a deficiency of local blood supply, it may be due to a deficiency in the diet [p. 177], or it may simply be due to a lack of vigour in the cells responsible for the repair work and a diminished capacity of the stem cells to produce more of them.

The thin atrophic skin of the elderly does not heal soundly, and the sluggish blood supply tends to allow the production of ulcers. The best example of this is afforded by the 'bed-sores' which occur at the sites on which pressure is taken by the skin of elderly people if vigorous preventive measures are not adopted. Experimental work suggests that, although the power to replace cells following a skin injury is greatest in infancy, there is little difference in the degree of cell replacement between adult and senile animals, though there is a measurable delay in reaching the peak rate of mitosis in old individuals. What is deficient is the strength of the scar; in old age collagen formation is delayed and the rate of contraction of the scar tissue is slowed. A major factor in the sometimes poor healing of wounds in the elderly is the fact that their diet is often deficient in both protein and vitamins [p. 175].

Similar findings are recorded when part of the liver is removed in experimental animals. Mitotic figures and binucleate cells decrease in frequency in sections of normal liver as age advances, but regeneration occurs in senile liver just about as well as in young adults, though here again the time before the mitotic count reaches its maximum increases in old age.

Not only do cells tend to lose their capacity to reproduce in old age, they also lose their ability to hypertrophy in response to stress; compensatory hypertrophy of one kidney in response to the removal of its fellow is relatively poor in elderly people. The recovery from nerve injuries [p. 180] becomes progressively less satisfactory as life lengthens. Probably the major cause of this is an inability to form new connexions within the central nervous system to compensate for misdirection of nerve fibres [p. 181], but an additional reason is a diminution in the ability of the damaged nerve cells to push out protoplasm to the periphery of the body.

CHANGES IN TISSUE MATRIX WITH AGEING

The matrix of a tissue largely determines such important properties of the tissue as its rigidity, elasticity, adhesiveness, and lubrication. In old age changes in the matrices of various tissues affect all these properties. Some have gone so far as to say that the difference between a unicellular organism, with no matrix, and a multicellular one with matrix between the cells, is the difference between potential immortality and inescapable death, since the matrix must age, and the alterations in it will topple the whole complicated edifice.

The first of these alterations is generally accepted to be a loss of fluid content, and this is usually well seen in connective tissue. In the embryo, connective tissue has a very high content of water [p. 4], but through life this gradually diminishes, and at the same time there is a gradual lessening of the amount of the amorphous material in the matrix and a relative increase in the fibrous elements it contains. The decrease in body water in old age appears to be more marked in women than in men.

The collagen fibres increase in number and also in thickness, and they develop more cross linkages with their fellows. In areas exposed to wear and tear they may become bent and damaged, and their staining reactions alter. The elastic fibres become less springy as they become thicker, and there is a tendency for them to fray and ultimately to fragment. A decrease in the extensibility of the elastic fibres in the aortic wall has been demonstrated.

The third important change is the deposition of calcium salts, particularly round the collagen fibres. This is very easily observed in the various types of cartilages. As age advances, the relatively structureless matrix of hyaline cartilage is invaded by collagen fibres and the tissue is thus converted into fibrocartilage; calcium salts are then deposited round the fibres, so producing calcified cartilage, and ultimately true bone may be formed. This process occurs quite early in adult life in the cartilages of the larynx (except, usually, for certain parts of the arytenoid cartilages, which remain elastic), and in the cartilages joining the anterior ends of the ribs to each other and to the sternum. In fibrocartilage the collagen fibres may thicken and clump together to form what is sometimes called 'asbestos'.

Another important site of calcium deposition is in the walls of the blood vessels. In an old person the arteries may contain a good deal of calcium, which converts the formerly elastic and contractile apparatus into rigid tubes. This has important functional consequences, such as an inability to respond to an emergency by expanding and forming a collateral circulation. It is thought that the deposition of calcium in the tissues hastens the disruption of the elastic fibres, and it also increases the viscosity of the tissue hyaluronic acid (p. 11).

The effects produced by these changes throughout the connective tissues of the body are manifest in a general diminution of elasticity and resilience. The skin loses its elasticity, and wrinkles appear; these are partly the result of changes in the arrangement of the bundles of collagen and of the elastic fibres in the dermis. The sagging and general loss of tension in the skin is due more to this cause than to alterations in the collagen or fluid content.

The stiffness in the joints which is such a characteristic feature of old age is largely caused by the loss of elasticity in the connective tissue round them, and this is mostly due to the deposition of collagen fibres. The transformation of the hyaline rib cartilages into calcified cartilage means that the process of breathing is interfered with. Elevation of the ribs depends largely on the power of these cartilages to twist on themselves, thereby allowing movement between the ribs and the sternum. If this is abolished by the cartilages becoming rigid, thoracic breathing is impeded [FIG. 43]. At the same time the chest is less able to withstand a crushing force applied to it than the chest of a young person, in whom the pliable cartilages 'give' a little instead of breaking.

The loss of fluid and the increase in the fibrous components of the tissues may have an effect on tissue regeneration and repair, since the movement of the active cells through the tissues is necessarily impeded by the increase in density and lack of resilience.

The elasticity of the lungs, which allows them to recoil during expiration, and is an important factor in the expulsion of air, is diminished in old age, with the result that the lungs do not empty themselves so efficiently. At the same time the vital capacity is reduced because of the fixation of the chest wall: after the age of 60 the male range is 2·4 to 4·7 litres compared with 3·5 to 5·9 litres between the ages of 20 and 40.

The loss of elasticity affects the stomach, which loses its capacity to dilate following a meal, and this is a source of dissatisfaction to such elderly people as retain a good appetite. In the colon there is a progressive increase in the number of diverticula, with a corresponding increase in the incidence of diverticulitis. The lumen of the appendix may be partly or completely obliterated by fusion of its walls and fibrosis; in many cases this may be the result of previous inflammation.

In the eye, the lens becomes less plastic and loses its power to accommodate [p. 212], so maintaining a more or less fixed focus. By the age of 70 over 90 per cent of people have some degree of opacity of the lens (senile cataract), possibly as a result of its continued growth [p. 82] being restrained by a tight capsule, so producing compression of the nucleus and the surrounding fibres, which eventually coalesce into a homogeneous mass. In another type of senile cataract fluid accumulates among the peripheral fibres, forcing them apart.

But perhaps the most important general result of the changes in the tissue matrix is in the circulatory system. Normally the aorta dilates as the heart forces blood into it, and the elastic rebound of its walls maintains the pressure in the arterial system while the heart relaxes and the aortic valve is shut. The diastolic blood pressure therefore depends very largely on the elastic rebound, and if in old age the aorta is converted into a more or less inexpansible tube, the result is that the blood pressure now falls sharply in between beats of the heart. The nutrition of the tissues is therefore somewhat impaired, since the steady flow of blood is converted into a more or less intermittent one.

Fibrous plaques appear in the mitral valve as early as the third decade; the chordae tendineae thicken, and so does the atrioventricular ring. Later, after about 60, calcification is common. The tricuspid valve is less severely affected, perhaps because it is subjected to less strain. At the same time the margins of the aortic and pulmonary valves become thicker and may calcify. Fat is deposited beneath the epicardium, sometimes becoming almost confluent over the whole of the surface of the heart, and infiltrates the myocardium of both atria and ventricles. The cumulative effect of these changes is to lessen the ability of the heart to respond to demands made upon it and to impede the working of the heart valves.

With advancing age, there is an apparent loss of muscle fibres in the sinu-atrial node, and in both the sinu-atrial and atrioventricular nodes the collagen, elastic, and reticular fibres become more prominent. In the atrioventricular node, infiltration by fat is a feature, sometimes as early as 30 years of age.

Gross morphological changes are detectable radiographically in the peripheral arterial system. Apart from calcification there is an increase in the tortuosity of vessels such as the arteries in the hand and foot, and as early as 18 years of age the vessels supplying the finger pads start to form a thick anastomotic network which becomes more complex with advancing age.

DEGENERATION

Degeneration is a difficult word to define: in general terms it means a breaking down of an organized structure to one less organized in form or in function, but this has different results in different tissues, and the process of degeneration in (say) bone is very different from the process in (say) blood vessels. Degeneration occurs at different times in different tissues and systems, and is closely linked with impairment of growth and repair.

Bone loses calcium in old age, particularly in women, and becomes porous and brittle. At the same time the diet of many old people becomes deficient in calcium, and their ability to absorb it declines. The term 'osteoporosis' describes the reduction in the volume of bone tissue per unit volume of anatomical bone, and between youth and old age this loss amounts to about 15 per cent of the skeleton, particularly of trabecular bone. The sequential disappearance of the trabecular pattern in the upper end of the femur [p. 65] is used as an index of the severity of the condition. At the same time there may be a decrease in the thickness of the cortex of the long bones and the vertebral bodies, and the Haversian canals often increase in diameter and become filled with fibrous or adipose tissue. Fractures are thus much more easily caused in elderly than in young adult patients, and sometimes bones may be broken in bed as the patient tries to turn round against the resistance of the bedclothes. By the age of 80 a woman has a one in five chance of sustaining a fracture of the neck of the femur. Vertebrae may collapse under the load of the body weight, with resultant loss of height and pain in the back.

Degenerative changes in cartilage include fatty degeneration of the cartilage cells, replacement of cells by matrix, and deposition of fat around the cells. Cysts may also be formed in the matrix.

The articular cartilages become thinner and they may actually be eroded through in places, so exposing to the stresses of movement the sensitive ends of the bones rather than the insensitive cartilage. The pain so produced increases the reluctance of the patient to move the joint, and this appears to encourage further deposition of collagen fibres round it, so that the joint may ultimately become stiff and painful. This is the condition known as senile osteoarthritis [FIG. 69], and is usually accompanied by the out-growth of bony processes known as osteophytes from the margins of the articular surface. These also impede movement and contribute to stiffness.

The fibrocartilaginous intervertebral discs degenerate after the age of 45 to 50 years, or even earlier, and tend to collapse under strain. There is a progressive reduction in the water content of the nucleus pulposus, followed by alterations in the protein—polysaccharide complex of the nucleus, and ultimately the whole disc becomes dehydrated, pigmented, and fibrotic. This causes a decrease in height of about 3 per cent, which is particularly marked in the lower part of the vertebral column, since this region takes the greatest gravitational load. The primary curvature in the thoracic region, which depends mainly on the shape of the bones, is relatively little affected, but both secondary curvatures are partly undone, and the posture becomes stooping. This is a reversion to the condition in the infant [FIG. 48], and is encouraged by the increasing weakness of the postural muscles

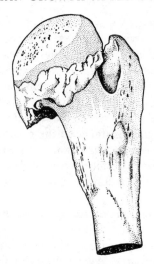

FIG. 69. Osteoarthritis. Showing marked erosion of the head of the
femur, and outgrowths of bone (exostoses) round the articular surface.
(Redrawn from a specimen provided by Dr. MacCallum and illustrated
in Boyd, W., *Pathology for the Surgeon* (1967) ed. Anderson, W. 8th ed.,
Philadelphia, W. B. Saunders Company, by kind permission of the editor
and publishers.)

which stretch across the secondary curvatures and help to maintain them.

Because of the compression of the lumbar discs, the xiphister-
num is brought nearer to the pubis, and the muscles of the
anterior abdominal wall are thus slackened off. At the same time
they are often required to cope with an accumulation of fat
within the abdominal cavity and in the abdominal wall itself. The
result is the characteristic bulging of the abdomen which occurs
after middle age.

It is difficult to say just how much of the observed loss in height
after middle age is due to degeneration and how much to the
slackness of the postural muscles. There have been very few
longitudinal studies of adult height, but a recent survey of adult
males in England and Wales showed that men of 60 to 64 years
of age are nowadays about 5 cm shorter than men of 20 to 25.
Much of this difference might be explained as a secular [p. 152]
rather than an age effect, and the difficulties in identifying and

isolating such trends by consideration of cross-sectional data have already been stressed [p. 133].

The heart muscle tends to degenerate in old age; it often becomes partly replaced by fat or fibrous tissue, and the result is an inefficient circulation. The poor blood supply to the brain combines with the loss of nerve cells to make manifest such symptoms and signs as a general loss of mental capacity—loss of memory, reasoning power, and concentration. The inability of the heart to provide the tissues with adequate supplies of oxygen leads to breathlessness on exertion, or eventually at rest, to coldness and blueness of the extremities, and perhaps even to gangrene of the lower limbs. Because of the impeded venous return [p. 139], the long columns of superficial veins in the legs and thighs often become varicose.

The heart has a further difficulty to contend with, for its own blood supply may become interfered with by degeneration of the coronary arteries. This is but one example of the generalized condition known as atherosclerosis, which affects the walls of the arterial system, particularly at points where mechanical stresses are greatest. Small plaques of fibrous and fatty material appear and may protrude into the lumen of the vessel, so encouraging the formation of clot. These patches in the wall weaken it, and may allow it to burst under the pressure of the blood inside, or to dilate like a balloon to form an aneurysm. The bursting of a vessel supplying the brain following a sudden increase of blood pressure is one way in which a 'stroke' can happen. Another way is by a clot becoming deposited on the degenerated patches and obliterating the lumen of the vessel, for brain tissue has a high oxygen demand and may not be able to survive until the development of a collateral circulation. This is also the case in the heart, and in both organs much of the territory supplied by the affected vessel will die and eventually be replaced—if the patient survives—by a fibrous scar.

The cause of atherosclerosis is still unknown, though several factors have been incriminated, among which are diet, smoking, lack of exercise, and an ambitious driving personality. In recent years the onset of atherosclerosis appears to be occurring earlier, so that it is now found in relatively young people: fatty deposits can be recognized in the wall of the aorta even in small children. Nevertheless, it has been for a long time one of the characteristic degenerative changes in old age, and virtually every male over 60 years of age has coronary atherosclerosis visible to the naked eye. There is considerable truth in the saying that a man 'is as old as his arteries'.

A fatty infiltration of the cornea, just inside the corneal margin, and forming eventually a more or less complete ring, is known as the arcus senilis. It is an almost inevitable accompaniment of old age; in one series of ophthalmic patients nobody over 50 was free from some degree of the condition, and in 40 per cent of males over 60 it was complete and well marked. Fortunately it does not appear to have any deleterious effect on vision, and its association with the condition of atherosclerosis is not now regarded as so significant as it was formerly.

THEORIES OF AGEING

Since the days of Francis Bacon there has been a great deal of speculation, which is still going on, about the possibility of finding a single 'key' cause of ageing. If such a key could be found, steps could be taken to change the lock, and so to prolong life indefinitely. Unfortunately matters do not seem as simple as that, but it is perhaps worth while to end by discussing briefly the current theories regarding ageing.

It has been suggested that the generalized deterioration of the body, manifesting itself in a progressive loss of vigour and an increased susceptibility to disease, may be due to loss of, or injury to, non-multiplying cells. The obvious scapegoat is the central nervous system, in which the inability of the cells to renew themselves is present from birth, and this theory includes the suggestion that senescence might be the result of degenerative changes occurring in the nervous system. The idea that ageing is under nervous control is attractive for several reasons, but it is virtually without factual support.

A second theory is that changes in the properties of multiplying cells are responsible for senescence. The idea is that this might come about because the somatic stem cells form an unstable family which changes in some way, perhaps by mutation, just as a cancer cell alters, so causing a lack of capacity to grow and to effect repairs which is the basis of all the changes of old age. The inability of cells to divide for more than about 40–40 generations may set a limit to the survival of tissues, which must then atrophy; if these tissues are vital ones, death must result. A variant of this hypothesis is that the DNA mechanism of replication becomes in some way faulty with many repetitions, so that ultimately the daughter cells are no longer faultless copies of their parents. The enzymes of the cell would then alter in amount or in structure, so causing disordered function and ultimately cell death. In support of this idea is the fact that chromosomal abnormalities in human leucocytes increase from a rate of about 4 per cent at the age of 10 years to a rate of 15 per cent at the age of 80.

Quite apart from histologically observable anomalies, it is probable that changes may take place in the accessibility of the genes on the chromatin, and it has been suggested that ageing is accompanied by a progressive suppression of the activity of genes. In contrast, 'rejuvenation' or 'de-differentiation' of cells represents the freeing and activation of previously quiescent genes.

It has been supposed that the common factor which accounts for the difference between animals which age and those which do not is that cells which are no longer able to divide have as a rule limited lives. One suggestion is that the process of cell division may give rise, as a kind of by-product, to certain enzymes or other chemicals which are necessary for the proper functioning of the daughter cells. If division becomes infrequent, cells may lose some essential component. But it has to be admitted that the potential life of a human nerve cell, on the basis of present information, is over 100 years, and there are other arguments against the hypothesis.

Certain organs such as the ovary have an intrinsic programmed life span, which may depend on stimulation by the pituitary or on the number of oocytes remaining. Old ovaries transplanted into young mice are not rejuvenated, but young ovaries transplanted into old mice continue to function in their new hosts as young ovaries. Young ovaries taken and stored at low temperatures can be given back to their now elderly (rat) owners and will restore oestrus.

The suggestion has been made that ageing in warm-blooded vertebrates might be due to their having a fixed adult size, and that the cessation of generalized growth in some way affects the power of the cells to regenerate and replace damage done by wear and tear. On this basis animals such as the sturgeon and the large tortoises, which are supposed to remain capable of growing almost indefinitely, should not show signs of ageing. Little work has been done in this direction, but it has been found that the survival curve of little fish called guppies, which also do not stop growing, does not appear to differ very much from those of small mammals.

Reproductive decline is a feature of old age in vertebrates, and attempts have been made to associate the changes of old age with the alterations in endocrine balance which occur at this time. However, the evidence so far does not support the idea that senescence is the result of a diminution in the secretion of any single hormone, and castration in either sex does not appear to shorten life or alter the process of ageing. The pituitary gland, in view of its close association with growth, and its known regulatory function in regard to other endocrines, is the prime suspect in such a theory, but in fact the best fit with mortality curves seems to be provided by the decline in the secretion of steroids from the suprarenal cortex.

Another suggestion is that the change which initiates ageing is the decline in the basal metabolic rate. So far there is little evidence for this. The possible effects of diet, and perhaps especially diet in youth, have recently aroused considerable interest. Rats fed a diet deficient in calories, but satisfactory in minerals, vitamins, and essential amino acids, remain immature for long periods [p. 149]. When such rats are given a normal diet they start to grow, provided the period of deprivation has not been excessive, and may then proceed to live for up to twice as long as the controls, largely because of a decrease in their liability to such illnesses as pneumonia and malignant tumours. Lesser degrees of restriction—say by intermittent starvation—produce proportionately lesser effects, perhaps a life span of about $1\frac{1}{2}$ times normal. The gain in life expectancy seems to be associated with the prolongation of immaturity, since rats similarly underfed after reaching maturity show only a slight gain in life span, which is probably attributable to the mere prevention of obesity. It is well established that rats fed a superabundant diet die sooner than controls fed a normal diet, and suffer earlier from degenerative conditions and neoplasia.

There is an obvious temptation to apply these results to man, but there are many serious obstacles. In the first place, rats are not men, and have a very different growth curve [p. 27]. Secondly, different effects are observed in different genetic strains of rats. Thirdly, malnutrition in man always involves deprivation of essential food materials as well as calories, and such malnutrition, whether in rats or in man, invariably shortens life [cf. p. 149]. It would not be easy, and it certainly would not be ethical, to duplicate the experiments on dietary restriction on human subjects, and until such experiments have been done there is little use speculating further.

On the other hand, it is only too easy to overfeed human children, and indeed a gigantic experiment in this direction appears to be in progress in those countries with a high standard of living, where modern infant foods, coupled with advertising and medical propaganda, combine to induce mothers to try to make their children as big and as bouncing as possible. It is a hard business fact of the insurance world that dietary excess and overweight in the adult are not good for life expectancy. For example, the Metropolitan Life Insurance Company showed in 1960 that if the body weight was 20 per cent above average the mortality rose by 31 per cent for males aged 15 to 69. The increase in mortality was very similar for females aged 40 to 69, but less (21 per cent) in those aged 15 to 39. Similar figures were reported from British sources in 1969, and the excess mortality is largely due to vascular disease and diabetes mellitus. Non-smokers weigh more than smokers, and those who give up smoking put on weight. Yet the dangers to health of smoking as a method of overcoming obesity are perhaps greater than those associated with the original condition. The best remedy is that suggested by Shakespeare: 'Leave gormandizing; know the grave doth gape for thee thrice wider than for other men.' As yet there is little evidence about the implications of the fact that in the United States in 1966 at least 10 per cent of children were thought to be overweight, but this has not stopped theorizing. It is suggested, for example, that the current rise in cardiovascular disease may be related to overfeeding in infancy, and retrospective studies estimate that about one-third of obese adults date their obesity from childhood. Put another way, 80 per cent of children who show evidence of obesity at school remain obese into adult life.

There is no doubt that children are maturing earlier than they did formerly, and the argument is that if rats live longer when they are kept immature, may not the present rapid maturation of children actually shorten their lives? It is believed that the secular trend in maturation is not a departure from normal, but rather a return to it [p. 155], and such evidence as exists does not suggest that constitutional precocious puberty shortens life. Nevertheless, it is clear that on a matter of such grave consequence more information is urgently needed, and as H. M. Sinclair has said, 'insufficient thought has been given to the most desirable rate of growth, which is not necessarily the maximum rate'. The optimal growth rate for animals has tentatively been defined by A. E. Needham as the rate which puts the least unnecessary stress on the metabolism of the animal, and he points out that few species grow at the maximum potential rate; it may therefore be a mistake to try to speed up natural growth.

A modification of the idea that ageing is due to changes in the multiplying cells is due to Burnet, and lays the blame on the lymphatic system. Mutation in the lymphocytes, which are concerned with the immunity mechanisms of the body, might result in a steady increase in the tendency for the body to reject as foreign protein the substance of its own tissues—behaviour which is known as the auto-immune reaction. Such a theory could account for many of the changes in old age, but more information is needed.

Finally, many people believe that the primary cause of old age is to be found in the alterations which occur in the matrix between the cells. Such alterations might of course be a secondary result of changes in the cells, for it is the cells which are responsible for the characteristics of the matrix. Deterioration of proteins with a slow rate of replacement, such as collagen, might be involved, and the increase in matrix round the cells [p. 221] might affect diffusion through the cell membranes.

These, then, are the present concepts of ageing, and it will be evident that theory has far outstripped the available information. What is needed now is an extensive programme of information-gathering, and the increasing numbers of old people in the population should afford an opportunity for this. It is nevertheless clear that the problems of old age are inextricably mixed up with the problems of growth, and that any light shed on the phenomena occurring at one end of life will immediately be relevant to investigations proceeding at the other. It was Montaigne who said 'Death is but an end to dying'; it might be just as true to say that death is but an end to growing.

Suggested Further Reading

BAYER, L. M., and BAYLEY, N. (1959). *Growth Diagnosis*, 241 pp., Chicago.

BIRREN, J. E., ed. (1959) *Handbook of Ageing and the Individual*, 939 pp., Chicago.

BOELL, E. J., ed. (1954) *Dynamics of Growth Processes*, 304 pp., Princeton.

BOURNE, G. H., ed. (1961) *Structural Aspects of Ageing*, 419 pp., London.

BOURNE, G. H., ed. (1971) *The Biochemistry and Physiology of Bone. Vol. 3. Development and Growth*, 2nd ed, 584 pp., New York.

BROZEK, J., ed. (1965) *Human Body Composition*, 311 pp., Oxford.

BULLOUGH, W. S. (1967) *The Evolution of Differentiation*, 206 pp., London.

CHEEK, D. B., ed. (1968) *Human Growth*, 781 pp., Philadelphia.

CLARK, W. E. LE GROS (1971) *The Tissues of the Body*, 6th edn, 434 pp., Oxford.

COMFORT, A. (1964) *Ageing; the Biology of Senescence*, 365 pp., London.

COMFORT, A. (1974) The position of ageing studies, *Mechanisms of Ageing and development*, 3, 1–31.

COWDRY, E. V., ed. (1952) *Problems of Ageing: Biological and Medical Aspects*, 3rd edn, ed. Lansing, A. I., 1061 pp., Baltimore.

CRELIN, E. S. (1973) *Functional Anatomy of the New-born*, 81 pp., New Haven.

FALKNER, F., ed. (1966) *Human Development*, 644 pp., Philadelphia.

FLOREY, H. W., ed. (1970) *General Pathology*, 4th edn, 1259 pp., London.

GESELL, A., and ILG, F. L. (1946) *The Child from Five to Ten*, 475 pp., London.

GOSS, R. J. (1964) *Adaptive Growth*, 360 pp., London.

GOSS, R. J. (1969) *Principles of Regeneration*, 287 pp., New York.

HALL, D. A. (1976) *The Ageing of Connective Tissue*, 204 pp., London.

HAM, A. W., and LEESON, T. S. (1974) *Histology*, 7th edn, 1006 pp., London.

HARRISON, G. A., WEINER, J. S., TANNER, J. M., and BARNICOT, N. A. (1964) *Human Biology*, 536 pp., Oxford.

HAYFLICK, L. (1958) Human cells and ageing, *Scientific American*, March, 32–7.

KROGMAN, W. M. (1949) The physical growth of the child: syllabus, *Yearbook of Physical Anthropology*, 5, 280–99.

McMinn, R. M. H. (1969) *Tissue Repair*, 432 pp., London.

Marshall, W. A. (1977) *Human Growth and its Disorders*, 179 pp., London.

Mason, A. S., ed. (1972) *Human Growth Hormone*, 201 pp., London.

Needham, A. E. (1964) *The Growth Process in Animals*, 520 pp., London.

Nowinski, W. W., ed. (1960) *Fundamental Aspects of Normal and Malignant Growth*, 1025 pp., Amsterdam.

Sheldon, W. H., and Stevens, S. S. (1942) *The Varieties of Temperament*, 520 pp., New York.

Sheldon, W. H., and Tucker, W. B. (1940) *The Varieties of Human Physique*, 347 pp., New York.

Tanner, J. M., ed. (1960) *Human Growth*, Vol. 3, Symposia of Society for Study of Human Biology, 128 pp., Oxford.

Tanner, J. M. (1962) *Growth at Adolescence*, 2nd edn, 325 pp., Oxford.

Tanner, J. M., Whitehouse, R. M., and Marshall, W. A. (1975) *Assessment of Skeletal Maturity and Prediction of Adult Height, T.W.Z. Method*, 99 pp., London.

The British Council (1976) *Human Malformations*, British Medical Bulletin *32* No. 1, 98 pp., London.

Thompson, D'Arcy W. (1942) *Growth and Form*, rev. edn, 1116 pp., London.

Weiss, P. (1949) Differential growth, in *The Chemistry and Physiology of Growth*, ed. Papart, A. K., 293 pp., Princeton. Reprinted 1971.

Wolstenholme, G. E. W., and Cameron, M. P., eds. (1955) *Ciba Foundation Colloquia on Ageing: I, General Aspects*, 254 pp., London.

Index

Numbers in bold type indicate main references